SUPERPOWER

SUPER-POWER

Australia's
low-carbon
opportunity

ROSS GARNAUT

LA TROBE
UNIVERSITY PRESS

IN CONJUNCTION WITH BLACK INC.

Published by La Trobe University Press
in conjunction with Black Inc.
Level 1, 221 Drummond Street
Carlton VIC 3053, Australia
enquiries@blackincbooks.com
www.blackincbooks.com
www.latrobeuniversitypress.com.au

La Trobe University plays an integral role in Australia's public intellectual life, and is
recognised globally for its research excellence and commitment to ideas and debate.
La Trobe University Press publishes books of high intellectual quality, aimed at general
readers. Titles range across the humanities and sciences, and are written by distinguished
and innovative scholars. La Trobe University Press books are produced in conjunction
with Black Inc., an independent Australian publishing house. The members of the LTUP
Editorial Board are Vice-Chancellor's Fellows Emeritus Professor Robert Manne and
Dr Elizabeth Finkel, and Morry Schwartz and Chris Feik of Black Inc.

9781760642099 (paperback)
9781743821176 (ebook)

A catalogue record for this
book is available from the
National Library of Australia

Cover design by Akiko Chan
Text design by Dennis Grauel
Typesetting by Akiko Chan

CONTENTS

To the Mutthi Mutthi, Paakantiji and Ngyimpaa people,
and to Tom, in whose country this story begins.

To Kai, Harvey, Leila, Luis, Eza, Sam and Isla, who close this story.
And to Indigenous Business Australia and its big role in the next.

A PERSONAL REFLECTION ON AUSTRALIA'S CLIMATE CHANGE ODYSSEY

In the first half of August 2019, my wife Jayne and I took a journey through the Murray–Darling Basin. We spent a couple of days at Lake Mungo, where the ice age overflow from the Lachlan once filled lakes and supported communities that left us with what may be the oldest record on earth of complex human mind and spirit. Then we turned up the Darling to Menindee, where Harrison made his pile when Pardon won the cup. Once the source of Broken Hill's water, and of renowned grapes and fruit, in the past year the Menindee Lakes have been famous only for a plague of dead fish. They are mainly dry. A new pipeline is raiding the Murray to supply Broken Hill. We cut across to Wilcannia, once a bustling port collecting wool and copper ores brought in by camels and bullocks for shipping down the river to Goolwa and Port Adelaide. Here was a little water, held back by the barrages, low and still. Finally, we crossed the Lachlan at Hillston and the Murrumbidgee near Griffith, and returned to the magnificent but weakening Murray once more. This was beautiful Australian country, rich with the human heritage of 50,000 years and 200 years.

For many Australians, their personal heritage lies in the Basin.

Jayne's father, Tom, lived in Wentworth and the lower Darling until he and many others in the bush rode to Melbourne on the news of war in 1914. That ride led him to a beach in Turkey, as the rising sun lit the coastal hills on 25 April 1915. A decade ago, a board in the Wentworth RSL club remembered Tom and his brother. This building without its memorabilia is now empty beside the Darling.

What value do we place on Australian heritage? The question kept coming back as we travelled the Darling. For about 30 kilometres the bottom of the riverbed is wet from the flowback from the Murray. But beyond that, the sand between the rows of grand old river gums is dry, except for scattered stagnant pools. By one of these lay the skeletal head of a Murray cod, its mouth gaping wide enough to swallow the largest carp whole. The dried flesh on the cod's back had been gnawed by wild pigs, which had waded into the shallow pond and dragged out the helpless survivor of eighty summers and a dozen droughts. Around a fire on the riverbank one night, we were told of plans for a quad bike ride along the Darling bed from just north of Wentworth to the arid Menindee Lakes.

Mike Sandiford, professor of geology at the University of Melbourne, is using new, satellite-based remote-sensing techniques to map the water contained in the structures below the surface in the Darling Basin. For millennia, water in occasional floods has filled porous sands and cavities and seeped out to maintain the life of plants and animals through the long dries. But now these occasional floods are treated as surplus, and held back for irrigation in the northern Darling and its tributary, the Barwon. Professor Sandiford foresees lower run-off from higher temperatures, reduced average rainfall and more insistent demands of irrigation interacting to contrive the desertification of the Basin. These factors are reproducing the fate of the Tigris and the Euphrates several thousand years ago, when the riverways that nurtured the beginnings of agriculture and human civilisation evaporated into today's Iraqi deserts.

But in one way, this tragedy is different from that of the original Garden of Eden. The Adams and Eves of Mesopotamia had not eaten of the tree of scientific knowledge; they knew not what they did.

But we do. The tragedy of the Murray–Darling is a consequence of denial, and of knowledge not being applied to public policy.

This was not always the Australian way. In June 2019, I spoke at Bob Hawke's memorial service of the great Australian prime minister's conviction that broadly shared knowledge was the foundation of good policy in a democracy. And just one day after returning from the Darling, I presented a memorial lecture for one of Australia's greatest public servants. John Crawford's essential contribution to Australian public life was his commitment to knowledge based on research as the starting point for sound policy development. It was through Crawford that I met Hawke in the late 1970s, while working on a report for Prime Minister Malcolm Fraser on the future of Australian industry. I was inducted into a great Australian tradition, of which Hawke and Crawford were the most accomplished proponents in their respective spheres.

But today, public policy based on marshalling knowledge through research and analysis, and then nurturing public understanding of the issues, seems a distant dream. That it is not contemporary reality is the essential problem behind the tragedies of the Murray–Darling Basin and of policy on climate change and the energy transition. (At an international level, that is also the fundamental problem of global trade and development.)

In the 2008 Garnaut Climate Change Review, I drew attention to the historic increase in world food prices in the first decade of the new century. And I highlighted that improving living standards in the populous countries of Asia would make this a great opportunity over a long period for Australian farmers and therefore all Australians – unless climate change at home damaged Australia's supply capacity.

Global climate change mitigation was needed, or Australia's farm capacity would be reduced. The atmospheric physics was showing that climate change would see the movement south of climate systems, and therefore the drying as well as warming of the southern latitudes where most of Australia's agricultural value lies. The irrigation output in the Murray–Darling could decline by 90 per cent if the world failed to act.

By 2019, new knowledge has reduced uncertainty without much changing these predicted consequences. We can now see the effects anticipated in 2008. Average temperatures across Australia so far this century are over a degree higher than in the first half of the twentieth century. We have reliable records of inflows into the Murray since 1892. After taking out the Snowy and inter-valley transfers, and the highly variable (currently absent) flows from the Darling, the average inflow in the past seven years has been a quarter below the first century of observation.

The controversial Murray–Darling Basin Plan does not take into account declining inflows as a result of climate change. It is unsettling now to read a CSIRO panel's description from 2011 of how the original Basin Plan dealt with climate change:

> MDBA [Murray–Darling Basin Authority] has modelled the likely impacts of climate change to 2030 on water availability and this modelling is robust. MDBA has not used this information in the determination of SDLs [sustainable diversion limits] for the proposed Basin Plan but rather has determined SDLs using only the historical climate and inflow sequences. The panel understands that this reflects a policy decision by MDBA ...[1]

After public outcry about the fish kills in the Darling below Menindee in 2018, the MDBA published a report in February 2019 on the effects of climate change. It noted that there had been five major blue-green algal bloom events in the past thirteen years. There had been four in the preceding sixty-five years. The report stated that lower rainfall in the southern areas of the Basin and higher temperatures were reducing stream flows into the rivers. After 48 millimetres per annum average run-off from 1961 to 1990, it was 27 millimetres in 1999 to 2008. 'The timing and magnitude of long-term climate changes remain uncertain and difficult to identify and measure separately from natural variability,' the MDBA wrote.[2]

Maybe. But maybe it's imprudent to use historical data without regard for climate change in calculating the amount of water available for allocation.

The Murray–Darling Basin Plan on which Commonwealth and state ministers agreed in 2012 was built, at best, on hope. At worst, on obfuscation. Either way, it contradicted scientific reality. And even this plan – for all its inadequacy – has not been implemented as designed. The Murray–Darling would be in better health if it had been honoured.

Yet the damage that climate change has wrought so far is of modest dimension compared with what will follow – even if the world takes decisive action immediately. And it is utterly trivial compared with what is to come if we fail to take decisive action. My 2008 Review demonstrated that we are the most vulnerable of the developed countries to damage from climate change.

In Paris in December 2015, all members of the United Nations agreed to hold temperature increases below 2°C and as close as possible to 1.5°C. The best that we can hope for now is holding increases globally to around 1.75°C. This could be achieved if the world moves decisively towards zero net emissions by 2050.

But temperatures over land will increase by more than the average over land and sea. An increase of 1.75°C for the whole world would mean more than 2°C for Australia – twice the increase that this year helped to bring bushfires in August to New South Wales and Queensland.

Such temperature increases would present Australia with a massive adaptation task. The internal disruption would be hard enough – with the Murray–Darling Basin just one of a hundred fateful challenges. The changes in our neighbourhood would probably be even harder for us to manage. The problems our neighbours in south and south-east Asia and the southwest Pacific would face would certainly and quickly become shared problems. The geostrategic own goal scored by Australia at the Pacific Island Forum leaders' meeting in August 2019 was a response to one of the smaller problems, but reminded us of our vulnerability.

A failure to act in Australia, accompanied by similar paralysis in other countries, would see our grandchildren living with temperature increases of around 4°C this century – and more beyond.

I have spent my life on the positive end of the Australian discussion of many international and domestic policy and development issues. That positive approach to what was possible in our democratic polity was mostly vindicated by the unfolding of history during the Australian reform era from 1983 to the beginnings of this century. But if the nation were to experience the consequences of a failure of effective global action on climate change, I fear that the challenge would be beyond contemporary Australian society. I fear that things would fall apart.

So is it all bad news? What we now know about the effect of increased concentrations of greenhouse gases in the atmosphere has broadly confirmed the conclusions I drew from the scientific research available for my 2008 and 2011 Reviews. But on the other hand, these Reviews greatly overestimated the cost of meeting ambitious reduction targets.

The good news is very good indeed for Australia, and especially for rural and provincial Australia. If we are wise, we can change the political story of climate policy in this nation. Quite a few Australians once argued that atmospheric physics is bunkum; or that there is no point in reducing greenhouse-gas emissions because others will not; or that it is too costly to reduce emissions, no matter how expensive the result of a failure to act. To those Australians, I can say: circumstances have changed.

It took me some years to realise the extent of that change. The Reviews from eight and eleven years ago touched upon the unusually high quality of Australian solar and wind energy and other renewable resources. They noted exceptional opportunities for growing biomass and capturing carbon in the landscape. They mentioned the possibility of great advantage. A chapter in each was devoted to carbon farming. But the references to exceptional opportunities were almost in passing.

After completing my official reports, I continued to take a close interest in Australian renewable resources and clever inventions that would help the transition to a low-carbon economy. I introduced leaders of many established Australian businesses to promising proposals for profitable investment in reducing emissions. Australian business is generally slow in innovation. Established non-competitive and anti-competitive arrangements are unusually rich by global standards, and disruption of them is unattractive. So I started to take up some of the proposals privately. I worked with partners in South Australia to develop ZEN Energy. Later we brought in British businessman Sanjeev Gupta as a partner after his purchase of the Whyalla Steelworks. ZEN and SIMEC ZEN Energy built acceptance in South Australia of the use of utility scale batteries to stabilise the power system; completed the development work on Australia's largest solar farm; and now supply the power needs of the South Australian government and the South Australian Chamber of Mines & Energy buyers' group. With old colleagues, I am working on some transformative power system developments beyond South Australia through Sunshot Energy.

Meanwhile, my work as an economist was tracking the rapid fall in costs of new technologies for reducing emissions in industry. By mid-2015, I was convinced that what in 2008 and 2011 I had perceived to be a possibility of modest dimension had become a high probability of immense economic gains. I gave a public lecture in June that year at the University of Adelaide: 'Australia as the Energy Superpower of the Low-Carbon World'.

In this book, I explain how my thinking has evolved from the earlier reviews – and why I now believe that if Australia rises to the challenge of climate change it will emerge as a global superpower in energy, low-carbon industry and absorption of carbon in the landscape.

I begin, in Chapter 2, by outlining recent developments in scientific knowledge on climate change.

In Chapter 3, I discuss how to assess the costs and benefits of Australia doing its fair share in a strong global effort. Here, changes in economic realities have altered earlier conclusions, although my methodology remains unchanged.

Chapters 4 to 7 explore the many benefits and opportunities of the good news about the lower cost of cutting emissions. Australia is richly endowed with resources that allow it to prosper from a global movement to zero net emissions. If we take early and strong action in ways that build upon our natural advantages, we will not suffer a decline in living standards in the near future in conventional economic terms as we move towards zero emissions. Now, much more than was anticipated a decade ago, we can be confident that we will be richer materially sooner rather than later, as well as very much richer in human and natural heritage, should we embrace a zero-emissions future.

The economic improvement has two main sources. First is the extraordinary fall in the cost of equipment for solar and wind energy and of storage to meet the challenge of intermittency. Per person, Australia has natural resources for renewable energy superior to any other developed country and far superior to our important economic partners in northeast Asia. Together with our strengths in mining, this makes us the natural home of processing mineral ores and some foodstuffs. Second is the immense opportunity for capturing and sequestering, at relatively low cost, atmospheric carbon in soils, pastures, woodlands, forests and plantations. Rewarding people and organisations that own and manage land with incentives equal to the true cost of carbon emissions would lead to sequestration in landscapes becoming a major rural industry. I said in 2011 that it could be a new rural industry as large as wool. That now seems to me to be a radical underestimate of the potential.

Technologies to produce and store zero-emissions energy and to sequester carbon in the landscape are highly capital-intensive. They have therefore received exceptional support from the historic fall in global interest rates over the past decade. This has reduced the cost of transition to zero emissions, at the same time as it has increased Australian advantages. (I discuss low interest rates and their effects in the Appendix to Chapter 3.)

These main drivers of lower costs are supported by Australia's exceptional capacity to produce biomass as a base for industry. This will be valuable as the world moves away from coal, oil and gas as raw materials for plastics and other chemical manufactures. It helps as well that we have unusually good opportunities for geosequestration of carbon dioxide wastes. In both cases, we could draw intensively on established Australian strengths in the biological, metallurgical and engineering sciences and in managing land and other natural resources.

There is more to say about the changing economics of a strong global mitigation effort. In 2008, comprehensive modelling of the costs and benefits of playing our part in the global transition suggested there would be a noticeable but manageable sacrifice of Australian current income until early in the second half of this century, but that average incomes would then regain lost ground and the gains would grow late this century and beyond.

Today, calculations using similar techniques would give different results. Australia playing its full part in effective global efforts to hold warming to 2°C or lower would now show economic gains instead of losses in early decades, and much larger gains later on. Whereas the modelling in 2008 suggested that Australia would import emissions-reduction credits, today I would expect Australia to become an exporter of emissions permits.

Australia should have a much stronger comparative advantage in energy-intensive minerals and agricultural processing in a zero-emissions world economy than it had in the fossil-energy past.

If Australia is to realise its immense opportunity in a zero-carbon world economy, it will require a different policy framework. But here there is also good news. The advantages of the low-carbon world are so great for Australia that we can make a strong start even with incomplete and weak policies that are consistent with established state and federal commitments. Policies to support the completion of the transition can be built in a political environment that has been changed by early success.

For renewable energy, we can build on considerable recent investment in solar and wind generation. State government policies in Victoria and Queensland will underwrite the early momentum, as the impetus from the Commonwealth's renewable energy target fades. It will become clear through the 2020s that drawing 50 per cent of electricity from renewable sources in these states and the country as a whole is simply a milestone on the path to more comprehensive transformation.

Three early policy developments are necessary. None contradicts established government policy. First, the regulatory system has to focus strongly on security and reliability of power when most electricity is drawn from intermittent renewable sources. Second, the regulatory system must support transformation of the transmission system, to allow huge expansion of supply from those regions with high-quality renewable energy resources. This is likely to require new mechanisms to support private initiative. Third, as a first step towards imposing order on a highly unstable and uncertain policy framework, the Commonwealth government could underwrite new investment in firm electricity supply, thereby securing a globally competitive cost of capital for a capital-intensive industry.[3]

For use of competitive power in expanding energy-intensive industry, grants for innovation in low-emissions industry along the lines provided by the Australian Renewable Energy Agency to renewable energy would also help. Among other things, this would support the hydrogen strategy being developed by Chief Scientist Alan Finkel.

In Chapter 4, I focus on an expanded role for renewable energy in electricity supply. Chapters 5 and 6 then discuss how this will take us a long way towards decarbonisation of transport and industry.

The full emergence of Australia as an energy superpower of the low-carbon world economy would encompass large-scale early-stage processing of Australian iron, aluminium and other minerals.

But for other countries to accept the shift to Australia of lower-cost low-emissions processing, and to import large volumes of low-emission products from us, we will have to accept and be seen as delivering on emissions-reduction targets that are consistent with the Paris objectives. Paris requires zero net emissions by mid-century. Developed countries have to reach zero emissions before then, so interim targets have to represent credible steps towards that conclusion. Japan, Korea, the European Union and the United Kingdom are the natural early markets. China will be critically important to realisation of the full opportunity. Indonesia and India and their neighbours in southeast and south Asia will sustain Australian exports of low-emissions products deep into the future. For the European Union, where carbon prices are now much higher than Australia's 'carbon tax' from 2012 to 2014, reliance on imports from Australia of zero-emissions aluminium, iron, silicon, ammonia and other products processed from energy and mineral ores would only follow assessments that we were making acceptable contributions to the global mitigation effort. We will not get to that place in one step or soon. But likely European restrictions on imports of high-carbon products, which will exempt those made with low emissions, will allow us a good shot.

The Australian emissions trading scheme was due to be integrated into the European one from 1 July 2014. Those arrangements went into hibernation with Australian carbon pricing. If something like them were brought back to life, we could now expect Australia to be a rapidly expanding exporter of goods embodying renewable energy, and to be engaged in close discussion of adjustments in rules to allow large-scale trade in legitimate carbon credits from the land sector. We can make a significant start on developing an important carbon farming industry through domestic markets, and go further when policy change allows large-scale international trade in carbon credits.

Alongside our strength in renewable energy, our advantages in growing, using and sequestering carbon in biomass will set Australia up as the international superpower of the low-carbon world economy. But for this to occur, Australia will need to regain its former strength in research and education on agricultural, pastoral, forestry and related industrial activities.

The low-carbon world economy will be especially favourable for rural and provincial Australia. Energy will be produced mainly outside the large cities, much of it in remote locations. This will make it commercially attractive to process many Australian mineral and agricultural goods into products of higher value close to the sources of the basic commodities. A new carbon-farming industry, prospering exceptionally in less agriculturally productive regions, will add substantially to rural incomes. Biomass will have additional value as a base for new industry, especially when combined with low-cost energy. The new farm- and station-based activities on average will make fewer demands on water than the old. And low-cost energy will improve the economics of recycling, desalinating and transporting limited water resources. Rural and provincial Australia will be the engine room of the superpower of the low-carbon world economy. Much of the new opportunity will be on land managed by Indigenous Australians.

Alas, the low-carbon opportunity cannot restore the life of the Murray cod on the dry bed of the Darling below Menindee. The unavoidable increases in carbon dioxide before we achieve zero emissions will keep on doing what carbon dioxide does. We will still leave for our grandchildren an awful job of cleaning up our mess.

Awful, but maybe not impossible. The low-carbon opportunity can make life better for the Australians who come after us. There is a better chance of leaving a manageable mess if we can build a bridge to a low-carbon economy, over which Australians can now walk to join the global effort on climate change. This book describes that bridge.

EXORCISING THE DIABOLICAL POLICY PROBLEM

In my 2008 Climate Change Review, I described climate change as a diabolical problem with a saving grace.

Climate change was harder than any other issue of high importance that had come before our polity in living memory. One of the reasons why it was hard politically was that it was not drawn in clear lines but blurred in uncertainty about its form and extent. It was also hard because its effects and remedies would play out over long time frames. The costs would come early and the benefits late, so that we had to think about how we value our own welfare alongside that of people who come after us. It was hard because effective remedies lay beyond any act of national will, requiring international cooperation of unprecedented dimension and complexity – with each country trusting that others would play their parts. The costs, of both damage and mitigation, varied both across and within countries. Finally, it was hard because vested interests were politically engaged and accustomed to the exercise of power.

I also noted in 2008 that daily debate in Australia and elsewhere suggested this issue may be too complex for rational policy-making.

The vested interests are too numerous, powerful and intense. The time during which effects will become evident is too long, and the time in which effective action has to be taken too short.

But there was a saving grace that might make all the difference: there was a much stronger base of support in Australia for reform on this issue than on any other large structural change that has come before our polity in recent decades. And people in other countries, to varying degrees, seemed to share Australians' concern about the issue.

How has the great struggle played out over the past decade – between the diabolical and the saving grace; between the vested private and the public interests?

In what follows, I tell the story of that struggle and seek to illuminate the path ahead. We will see that in the world as a whole the struggle between the diabolical and the saving grace has gradually, haltingly and incompletely, but I think inexorably, turned in favour of action on climate change. And that Australia would prosper exceptionally from doing its fair share in a strong global effort to reduce the disruption from climate change.

But in Australia itself, the outcome is not yet so clear. Polls reveal that there is now strong – and increasing – support for action on climate change. However, the electoral arithmetic is complex and muddied by coalmining becoming a proxy for climate change.

Climate change and economic development

Climate change will not be stopped by ending development. The challenge is to change the relationship between economic growth and emissions of the greenhouse gases that cause climate change.

Modern economic development was built on fossil fuels. Solar energy and atmospheric carbon dioxide had been converted by photosynthesis and natural storage processes into coal, oil, gas and other carbon compounds over many hundreds of millions of years.

That process had created the atmosphere and the climate that made our type of life possible. With modern economic development, these reserves of fossil energy have been drawn down rapidly, to drive the machines of the newly industrial world and meet the expanding demands of households. Coal dominated the mix of fossil fuels at first, and the balance shifted towards oil early and natural gas late in the twentieth century.

Knowledge of the physics of climate change brings forward what was always an inevitable need for transition from fossil fuels to other forms of energy. We are learning that the transition is possible without sacrificing living standards in currently wealthy countries or disappointing hopes for improving conditions in the developing world.

The extension of sustained modern economic growth into China and the other populous countries of Asia in the late twentieth and early twenty-first centuries placed huge pressure on easily accessible fossil energy resources. Prices rose six-fold in the decade to 2012.

The China resources boom raised anxieties about the cost of fossil energy resources for extending modern economic growth to the whole of humanity. But as it has turned out, climate is the binding constraint – not fossil energy reserves. If we can reconcile zero global carbon dioxide emissions with a continued rise in global living standards, this will remove the fossil energy constraint as well as preserve the climate.

Our shared scientific understanding

Over the past decade, we have seen considerable expansion of scientific knowledge about climate change. This has generally reduced uncertainty without fundamentally changing our understanding of the impact of greenhouse-gas concentrations on the climate.

The science is genuinely complex, and so in this book I have sought guidance from experts.

First, our knowledge of climate sensitivity – the amount of warming associated with a doubling of greenhouse-gas concentrations in the atmosphere – has expanded somewhat. In 2008, I took the position that the global average temperature could be expected (in probabilistic terms) to increase by about 3°C for each doubling. At that time, there was a great deal of uncertainty around the median expectation. The presence of aerosols and a range of feedback effects meant that it would take a long time – more than a century – for the full effects to be felt. Research over the past decade has confirmed this view of the most likely outcome, and reduced – without removing – uncertainty. Scientific research continues on long-term and short-term feedbacks, as well as different components of radiative forcing. Put briefly, carbon-cycle feedbacks that are not yet in the standard Intergovernmental Panel on Climate Change (IPCC) assessments are estimated to increase warming by about 0.25°C for a 2°C forcing. The emissions reductions that were associated with a 1.5°C outcome in earlier IPCC reports will probably only hold warming to 1.75°C. When one is considering the difference between the Paris 1.5°C and 2°C targets, an additional 0.25°C of warming becomes significant.

Much more knowledge is being generated about the role of both ice–albedo and carbon cycle feedbacks in the climate system. These feedbacks, although not large individually at 1.5°C or 2°C forcing, add up to a significant effect on remaining carbon budgets. The passing of time has seen changes in temperature, sea levels and other impacts broadly in line with expectations.

Since the 1950s, there has been a consistent pattern of Australian average temperatures being higher in each successive decade – and notably higher from the 1970s (see Chart 2.1).

Increases in global average sea levels have continued to press at the top of the range anticipated by scientific reports available in 2011 (the shaded area in Chart 2.2).

Chart 2.1 Australian average annual temperatures 1910–2018

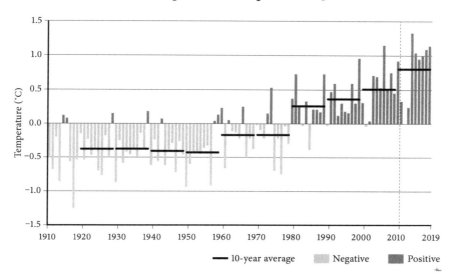

Note: The data show temperature difference from the 1961–1990 average.
Source: Bureau of Meteorology time series data, retrieved 29 March 2019.

Chart 2.2 Global mean sea level rise 1993–2019

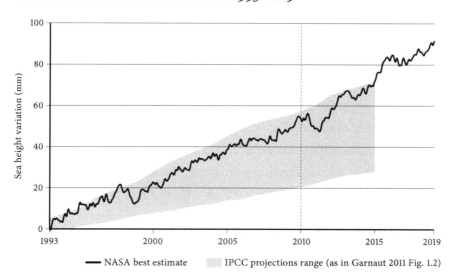

Source: NASA Goddard Space Flight Centre, accessed 11 September 2019; Garnaut 2011.

Since 2010, we have increased our knowledge of sea-level rise, largely due to improved understanding of how Antarctic and Greenland ice will respond to climate change. If emissions stay high, it is now thought that the 2100 global-mean sea-level rise could be 50 centimetres greater than previously projected. This understanding also widens the range of responses, with the new higher end for 2100 being about 90 centimetres greater than previously estimated. However, in low-emissions scenarios, there is a more uncertain story: the range of responses is wider, but the best estimate does not move very much at all. The contemporary science tells us that these processes are likely to lead to the risk of a very large sea-level rise if strong mitigation action is not taken, but that this can be mostly avoided if we move swiftly to establish zero net emissions over the next few decades.

I noted in my 2011 Review that one of the early footprints of anthropogenic climate change was drying and warming in Mediterranean regions. I considered the data on streamflow into dams in the Perth region of Western Australia. There has been another step downwards in streamflow over the past decade (Chart 2.3), rendering that city of over two million people overwhelmingly dependent for household, commercial and industrial water on desalination and declining groundwater.

The 2008 Review looked closely at risks to the Murray, without presenting projections from the scientific modelling. The data reveal a stepping down of inflows into the Murray over recent decades. Detailed records are available since 1892. In the last twenty years, to 2018, the average inflow has been 5826 gigalitres per annum. In the previous 107 years, the average was 9628 gigalitres. The tendencies are consistent with the grim storyline from 2008.[1]

Experience and advances in scientific knowledge have led to a greater capacity to attribute the probability of extreme weather events to climate change. The approach has been risk-based – similar to that used, for example, in attributing lung cancer to smoking. Scientists

Chart 2.3 Annual stream inflow to Perth dams

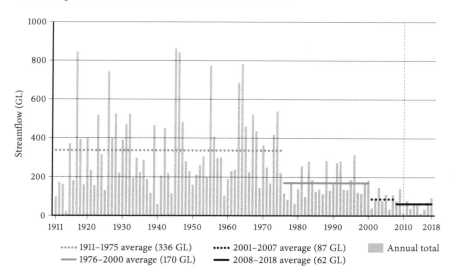

Note: Values exclude Stirling and Samson dams.
Source: Water Corporation (Western Australia), accessed 29 March 2019.

have observed that many costly extreme weather episodes in recent years have been rendered much more likely because of climate change that has already occurred.

The recent science has looked more closely at the implications of all greenhouse gases not being the same. There is more explicit recognition that zero net emissions can be achieved through a combination of zero emissions for long-lived gases (carbon dioxide) and stable emissions for short-lived gases (methane).

What are the implications of all these developments in science? The passing of time has led to a stronger focus on the damage from 2°C of warming, and the advantages of holding temperature increases well below that level. At the 2015 Paris meeting, United Nations members agreed to hold emissions below 2°C and as close as possible to 1.5°C. As we have seen, the numbers used by the IPCC mean that the '1.5°C' outcome will really be 1.75°C. A 1.5°C carbon budget allows for fifteen years of today's emissions, so we have about thirty years to get to zero if we proceed in a straight line to that target.

Once we have set a 'target' for temperature increases, we can calculate the 'budget' for greenhouse gases that can safely be emitted into the atmosphere while still allowing us to reach that target. If the world makes a late and slow start on reducing emissions, it uses up most of the budget early. If we had moved with resolve from 2008, we could have reduced to zero over about fifty years. Of course, if we continue to move slowly – as we have so far – we will have to achieve zero emissions earlier than 2050, as it is the accumulated total of emissions that determines the warming effect.

Zero net emissions means zero emissions of gases that remain in the atmosphere over long periods (most importantly carbon dioxide, with an average 'residence' of thousands of years), and stable emissions of gases that have short lives in the atmosphere (for example, methane). We don't actually have to reach zero for methane emissions – although the closer we get to that level, the more likely it is that we will achieve the overall goal of net zero emissions at an early date.

These, of course, are global targets and global budgets. Generally accepted ethical principles, and the governments of developed countries in UN negotiations, support the view that developed countries need to reduce emissions more rapidly than developing.

The ethics of climate change

Human ethical systems grew out of social experience. Climate change takes us beyond the historical experience. It is inevitable that we have taken time to sort out the ethical foundations for this new challenge.

We know beyond reasonable doubt that our participation in unconstrained consumption and production in the old style of modern economic development will damage the lives of others through climate change. Many of the people who are harmed will be in other countries, and in the future, and in any case we cannot be sure who will be affected by our actions. Does this make failure to act on climate change, collectively and individually, less reprehensible? The application of widely accepted ethical principles leads to the clear conclusions that it doesn't.

The most rigorous, comprehensive and influential treatment of the ethics of climate change is Pope Francis's 2015 encyclical *Laudato si'*. In this work, he applies Catholic, Christian and general ethical teachings and intellectual traditions to climate change. At one level, Francis speaks to the third of humanity who, at least in polls, profess the Christian faith (and to some extent also the quarter who profess other Abrahamic faiths). At another, he explains how people from all ethical systems have reason to take strong action to combat climate change.

The encyclical is grounded in sound contemporary knowledge of atmospheric physics:

A very solid scientific consensus indicates that we are presently witnessing a disturbing warming of the climate system ... most global warming in recent decades is due to the great concentrations of greenhouse gases ... released mainly as a result of human activity ... The problem is aggravated by a model of development based on the intensive use of fossil fuels ... If present trends continue, this century may well witness extraordinary climate change and an unprecedented destruction of ecosystems, with serious consequences for all of us ... There is an urgent need to develop policies so that, in the next few years, the emission of carbon dioxide and other highly polluting gases can be drastically reduced, for example, substituting for fossil fuels and developing sources of renewable energy.[2]

Laudato si' sees climate change imposing costs disproportionately on people in poor countries and poor people in all countries, and thus as a social justice issue. Francis declares that it is a flaw in our political and economic systems that the private profit motive unconstrained by concern for the public good has led us to damage our common home – the same flaw that has generated and tolerated great poverty.

Laudato si' argues that humans have a responsibility to take care of the natural environment, both as something important to humanity and as something valuable in itself. Francis interprets humans' 'dominion over the earth'[3] as a responsibility to 'protect the earth and to ensure its fruitfulness for coming generations'.[4]

John Broome, professor of moral philosophy at Oxford University, has reached similar conclusions from secular ethical principles. He reached the strong conclusion that ethical behaviour requires each individual to take responsibility for his or her own emissions by eliminating them or offsetting them. Each tonne of greenhouse-gas emissions is reasonably expected to impose some extra damage on someone, somewhere. That the victim is unknown does not diminish the responsibility. This personal ethical obligation is in addition to our responsibility as citizens to work for the introduction and implementation of policies that achieve good climate outcomes.

Developments in global action

Since 2008, we have seen a fundamental change in the framework of international cooperation, which has improved prospects for global action.

In 2008, the challenge was what I called a 'prisoners' dilemma'. It was costly for each country to reduce its own emissions; all countries together would benefit from all countries reducing emissions to levels that added up to acceptable limits on global warming; each country, if left to itself, would do too little, leading to underperformance against global objectives; and a binding global agreement was necessary for effective mitigation.

A binding agreement had the additional virtue of supporting international trade in emissions permits. Such trade would reduce the cost of meeting targets in both surplus and deficit countries. A global price for carbon would remove distortions in investment in emissions-intensive industries. That international price would be in the vicinity of $40 per tonne in 2008 prices to achieve a 2°C target.

My 2008 Review went beyond the conventional wisdom under-pinning international climate policy in revising upwards expectations of business-as-usual emissions from China and other rapidly grow-ing economies to reflect the early-twenty-first-century reality. The clear implication was that China and other major developing countries would need to constrain emissions from an early date to avoid global disruption.

International agreement encompassing developing countries required a conceptual basis for dividing responsibility among countries for reducing emissions that was practical and widely regarded as being fair. I proposed what I called 'modified contraction and convergence'.

Within 'modified contraction and convergence', each developed country would commit to reducing emissions linearly from current levels to zero over a timeframe that, when considered alongside others' emissions, would hold global emissions within the budget. Developing countries would commit to steadily reducing the emissions intensity of their economies until such time as their total emissions had reached the declining average level of developed countries. From that point, the developing country's per capita emissions would decline absolutely in line with the developed countries.

This model was applied to recommendations on Australia's targets: a 25 per cent reduction by 2020 if the world as a whole entered com-mitments that would hold temperature increases to 2°C; 10 per cent for international commitments consistent with 3°C; and an unconditional commitment to 5 per cent if there were no international agreement at all. These recommendations were broadly accepted by the Australian government. With the explicit support of the Opposition under Malcolm Turnbull's leadership, they were tabled with the UN in advance of the Copenhagen meeting in 2009. And with the explicit support of the Opposition under Tony Abbott, they became Australia's pledge at Cancun a year later.

Copenhagen was a diplomatic fiasco, leading to widespread disillusionment with efforts to achieve international agreement through the United Nations Framework Convention on Climate Change (UNFCCC). But an alternative approach emerged in the embers of the Copenhagen meeting. Heads of government from the United States, China, Brazil, India and South Africa met to discuss what could be salvaged. At the time, I called the alternative approach the 'Obama Accord'.

The Obama Accord did not seek a legally binding agreement. Developing as well as developed countries would pledge voluntarily to reduce emissions. It was recognised that initial pledges would not add up to a satisfactory outcome. There would be successive UN meetings to review national pledges and exert pressure on laggards. The 'pledge and review' element of the new approach was suggested by the Australian delegation at Copenhagen.

The alternative approach was formally accepted and initial pledges were made at Cancun in late 2010. In practice, governments were prepared to make larger commitments when they were voluntary than when they were legally binding. (In this, international climate policy had much in common with trade policy. On trade, countries in Australia's western Pacific region from the mid-1980s to the late 1990s had gone much further and faster in reducing trade barriers within a framework of 'concerted unilateral liberalisation' than they had when negotiating binding bilateral and regional preferential trade agreements in the twenty-first century.)

I called the new approach 'concerted unilateral mitigation'. This approach also had the decisive practical advantage that it was consistent with US constitutional and political realities.

Neither the advantages of a binding agreement nor the disadvantages of concerted unilateral mitigation were as great as had been supposed, including by me. A 'binding' international agreement was not really binding. It could not be legally enforced in practice: the Canadian government in 2012 walked away with impunity from breaches of its Kyoto commitments. A domestic political commitment to a target was taken seriously and in practice mostly met or exceeded. This was the experience with the Cancun and later Paris pledges of the two biggest emitters, the United States (before Trump and at least in the first two years of his term) and China. And while legally binding targets would provide a firmer basis for international trade in 'surpluses' and 'deficits', voluntary targets taken seriously and seen by international partners to be satisfactorily ambitious could serve a similar purpose in practice.

The Cancun approach became the basis of the December 2015 Paris Agreement. All countries except Nicaragua and Syria were parties to it (subject to ratification through national processes). Nicaragua and Syria subsequently agreed to participate. The voluntary nature of commitments freed the US government from congressional veto.

These national commitments added up to a major change in trajectory for global emissions. China for the first time accepted absolute constraints on emissions – albeit with no commitment to peaking before 2030. The United States committed to a 28 per cent reduction by 2025 from 2005 levels – 2025 being the date which President George W. Bush had said in 2007 would mark the peak level of US emissions. Australia's pledge was among the weakest of those offered by developed countries – 26 to 28 per cent by 2030 from 2005 levels.

While embodying a marked strengthening of earlier commitments, the Paris pledges did not add up to achievement of anything like the 2°C, let alone the 1.5°C, objective. That was to be corrected over time. The temperature commitments were the primary ones, and numerical emissions-reduction targets would be adjusted over time to be consistent with them. Numerical targets would be strengthened periodically in subsequent 'pledge and review' meetings of the UNFCCC.

Most countries have performed reasonably well against the inadequate 2030 targets. China, in particular, was moving faster than promised at Paris. There was a marked change in trajectory in China from 2012. This has greatly enhanced the prospects for the world meeting ambitious mitigation targets (see Chart 2.4).

Regrettably, after falling for several years, carbon emissions in China rose again from 2017. The economic and environmental policy reforms under China's new model of economic growth from 2012 have been partially reversed since 2017, as China has sought to increase old-style economic growth to compensate for losses from the trade war with the United States. It is critically important to the global effort that China's emissions begin to fall again soon.

China's consumption of coal for electricity rose at double-digit rates from the beginning of the century to 2011. Chinese coal use was the biggest single factor behind the large increase in global emissions through this period. From 2013 to 2016 more than the whole of Chinese electricity supply growth came from hydro-electricity, nuclear, wind and solar sources. Solar grew most rapidly from a low base. While zero-emissions sources of power continued to grow strongly after 2016, the reassertion of elements of the pre-2012 economic model saw total electricity output accelerate, and a return to expansion in coal-based power generation, albeit at less than half the rate of the first decade or so of the century.

In 2017, the Trump administration initiated a process of withdrawal from the Paris Agreement. Meanwhile, US emissions have continued

Chart 2.4 China emissions: actual, old business-as-usual and Cancun commitment

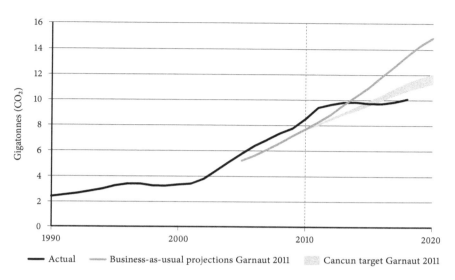

to fall, under the continued influence of pre-Trump regulatory action of the Federal Environmental Protection Agency, state governments, community pressure, falling costs in zero-emissions industry and the replacement of coal in electricity generation and many industrial uses by cheaper natural gas and renewable energy.

The US withdrawal process is unlikely to be completed during the current presidential term. Much hangs on the approach to international climate policy of the next president.

Four developments have supported nations in reducing emissions. In all countries, there is stronger community awareness of the seriousness of the challenge from climate change to established patterns of life. People everywhere have become more aware that countries other than their own are taking action, so that concern about others free-riding has diminished. The costs of emissions-reduction technologies have fallen dramatically – partly in response to expansion of scale from early action on climate change in Europe, generating an industrial response in China, and partly from a decline in interest rates and the supply price of investment. And there is increasing awareness of co-benefits of reducing emissions: competitive advantages from being early in applying new technologies; lower energy costs in countries with exceptionally rich endowments of renewable energy resources; health benefits from reduced atmospheric and other pollution from combustion of fossil energy; and sustainability benefits from retention of carbon in soils, pastures, woodlands and forests.

The Australian policy tangle

The Garnaut Climate Change Review was commissioned by the six state and two territory governments in April 2007. After the November federal election, the Commonwealth government led by Prime Minister Kevin Rudd joined the process. When I reported to the prime minister and state premiers at the end of September 2008, the main analysis and recommendations had the strong support of the federal Coalition Opposition.

Things fell apart in Australia in December 2009 when, on the eve of the Copenhagen conference, Tony Abbott won the leadership of the Liberal Party from Malcolm Turnbull by a single vote on the issue of whether the Opposition in the Senate would support legislation for an emissions trading scheme that had passed the House of Representatives.

Immediately after the 2010 election, I was commissioned to update my Review. I reported this time to the Multi-Party Parliamentary Committee on Climate Change. The main recommendations were legislated in late 2011 and early 2012 within the Gillard government's Clean Energy Future package. The legislation established carbon pricing, operating alongside the renewable energy target (RET) brought in by the Howard government and then strengthened under Rudd. That same package established the Australian Renewable Energy Agency (ARENA), the Clean Energy Finance Corporation (CEFC), the Carbon Farming Initiative (CFI) and the Climate Change Authority (CCA).

For two years from 1 July 2012, Australia worked within a broadly based, administratively smooth, environmentally effective, economically efficient and equitable set of emissions-reduction policies, largely governed by market processes. Each large-scale retailer or user of power had to relinquish to the Clean Energy Regulator its 'share' of renewable energy certificates, establishing a market for them. A carbon price covered electricity generation and industry representing about three-fifths of national emissions. The agricultural and land sectors, accounting for about one-fifth of emissions, and some other activities were brought within the influence of carbon pricing by the CFI, through which emissions reductions certified by the Clean Energy Regulator could be allocated Australian Carbon Credit Units (ACCUs). These credits could be sold to entities covered by the carbon price, which could use them to acquit their liability.

The carbon price was fixed for three years at $23 per tonne, rising at 4 percentage points per annum plus the percentage increase in the consumer price index. The fixed price was to metamorphose into a floating one within an emissions trading scheme when the Australian carbon-pricing scheme was linked to the European one from 1 July 2015 (later 2014). From that time, the carbon price would be set by the interaction of the Australian and European markets. Market interactions would cause the carbon price immediately to reduce the price of RET certificates, and over a number of years render the RET redundant.

Consistently with my recommendations, free permits were issued to trade-exposed businesses. Allocations were based on emissions levels before the introduction of the scheme and gradually fell over time. The system for allocating free permits established powerful incentives for the companies receiving them to reduce emissions. Many responded positively to those incentives. Inconsistently with my recommendations, free permits were provided to companies which owned the most emissions-intensive coal-based power generators – albeit at levels well below those recommended by the Australian Energy Markets Commission (AEMC). The AEMC said that the payments to emissions-intensive generators were necessary because in their absence their insolvency could cause power system failures. The coal generators were expected on average to pass nearly all of the carbon price on to users of electricity – and did so. They received the value of the free permits as a lump sum at the beginning of the scheme, and kept the proceeds when the carbon price was abolished in 2014.

The bulk of the permits were sold. This yielded around $7 billion per annum of revenue in the two years in which the scheme operated. The largest part of this revenue was used to reduce income tax and increase social security payments to low- and middle-income households in ways that encouraged participation in the labour force. This more than compensated low- and middle-income earners for price increases. The balance of the permit revenue was applied to funding the ARENA, the CEFC and the CFI.

The government made little attempt to explain the link between the increased cost of living from carbon pricing and the reductions in income tax and increases in social security payments. One view among political strategists for the government saw more political advantage in taking credit for the income tax and social security changes and saying as little as possible about carbon pricing. That did not help the political standing of the government's policy on climate change.

The introduction of the carbon pricing scheme by the Commonwealth Public Service proceeded remarkably smoothly. There was none of the teething pain that accompanied introduction of the goods and services tax in 2001. There was no disruption of or setback to industry.

Emissions fell in the ways and amounts anticipated. Business started to get on with the adjustment to a low-carbon future.

The decisive victory of the Liberal–National Coalition led by Tony Abbott at the 2013 federal election was interpreted as electoral rejection of carbon pricing. The evidence is, at most, equivocal. Opinion polls immediately before the 2013 election showed greater support for the government's policy of carbon pricing than the Coalition's 'Direct Action Plan'. Support for carbon pricing was even stronger when respondents were reminded about the use of the revenue to reduce income tax and increase social security payments and to support expansion of renewable energy. An exit poll conducted by JWS Research for the Climate Institute found that 40 per cent of Coalition voters thought it more important to meet the emissions targets than to repeal the carbon tax, and only 32 per cent thought that it was more important to repeal the tax.[5]

The Abbott government came to office committed to repealing legislation supporting the carbon price, the ARENA, the CEFC and the CCA. Once in office, the Abbott government added abolition of the RET to its repeal list. In a remarkable episode, the Palmer United Party, briefly holding the balance of power in the Senate, and initially committed to support the Abbott government's repeal program, used its numbers to block repeal of the Abbott legislation, excepting only carbon pricing. Its leader, Clive Palmer, explained in a joint press conference with former US vice president Al Gore at Parliament House that he had changed his mind about climate science. The RET was eventually substantially weakened, with the Opposition supporting the Coalition in the Senate.

The changes in 2014 have left a deeply problematic legacy. Abbott had promised in the 2013 election campaign to keep the income tax cuts and increased social security payments while abolishing the revenue base that had paid for them. The resulting budgetary problems helped to undermine the political standing of the Abbott government.

And it left an incoherent climate and energy policy legacy. Uncertainty in energy policy has contributed to electricity prices being much higher than when the carbon price was operating – with no government revenue to compensate low- and middle-income households.

I finished writing my book *Dog Days* in the week after the election of the Abbott government in 2013. I said that repeal of carbon pricing would not be the end of climate change mitigation. It would just increase uncertainty and the cost of action and reduce its effectiveness:

> The new government is bound by its election commitments to introduce legislation to remove carbon pricing. That legislation will pass the House of Representatives. If the legislation were to succeed in the Senate, it would deepen the budgetary problems with which the government will eventually have to deal. It would lead to larger sacrifices of productivity than would be necessary with broadly based carbon pricing. It would lead either to much greater costs later in the decade or to Australia breaching its commitments to the international community and damaging its own interest in the mitigation effort. And it would set the Australian polity on another long journey to find a way to make our contribution to combating climate change, distracting the government and the polity from the great economic challenges facing Australia.[6]

The rest is history.

RISK, REWARD AND THE ECONOMICS OF CLIMATE CHANGE

My 2008 Review set out an economic framework for taking decisions on whether and how much we should seek to constrain climate change. The framework was applied to decisions from an Australian national perspective. I compared the costs and benefits to Australia of several levels of climate change mitigation by drawing on elaborate studies. I concluded that the benefits of full Australian participation in a global effort to contain temperature increases to 3°C exceeded the costs. Benefits would exceed costs by a greater margin for 2°C.

At the time, these levels of ambition were at the higher end of the range that had been contemplated in serious economic studies. In his *Review on the Economics of Climate Change*, Nicholas Stern opted for 3°C (on the grounds that it was likely to be the best achievable outcome). William Nordhaus opted for higher than that because, he said, costs exceeded benefits at 3°C. Nordhaus noted that his difference from Stern on the optimal level of mitigation derived mainly from the higher rate at which he discounted future benefits.

I did not look at more ambitious mitigation objectives than 2°C, except to note that the achievement of 1.5°C would depend on early demonstration that 2°C was possible. For a 3°C objective, my study showed, the benefits of action exceeded costs even if we only focused on narrowly economic benefits, excluded all impacts on natural and human heritage values, excluded the insurance value of avoiding uncertain but highly costly outcomes, and excluded all benefits after 2100.

The costs of a 2°C outcome were just about fully matched by narrowly economic benefits this century. The narrowly economic gains beyond this century, and all of the insurance and non-economic gains, were much greater for 2°C than 3°C. A rigorous application of my framework may have supported 1.5°C at the time, but I did not undertake that analysis. I have little doubt that the framework would support 1.5°C with the much lower cost of low-emissions technologies today.

This chapter outlines the 2008 framework and conclusions, and their update in 2011. It then examines how the 2008 and 2011 conclusions have been affected by developments over the past decade in economic analysis, economic realities, and the scientific, ethical and political economy context in which the economic analysis is applied. It comments on the role of the discount rate and the effects of an historic fall in real interest rates.

Decisions by one country in a many-country world

Economists before my two Climate Change Reviews had sought to determine whether and the extent to which mitigation made sense for the world as a whole.[1] William Nordhaus received the 2018 Nobel Prize in Economic Sciences for his work in this field.

The Garnaut Reviews sought to answer the question of whether and the extent to which mitigation made sense for one country: Australia. Whether mitigation is justified for the world as a whole turns out to be an easier question than whether it is justified for a single country. An assessment for an individual country must deal with all of the complexities that others had addressed for the world as a whole, plus one. The additional one is the interaction of one country's efforts with those of the rest of the world. That additional source of complexity is perhaps the most difficult of all.

The relevant mitigation is global. A single country's action is relevant for its direct and indirect contribution to global mitigation. The costs of various levels of mitigation for a single country depend mainly on the extent of its own mitigation – although these costs would be substantially lower with a global agreement in which at least major economies pursued similar goals with compatible policies. Each country's evaluation of whether some mitigation action of its own is justified depends on its assessment of the interaction between its own decisions and those of others. Thus, its own decision-making framework must incorporate its assessment of the dynamics of complex interactions among many countries.

The global mitigation effort is the sum of the separate but interrelated decisions of individual sovereign countries. In 2008, I discussed how what one country did would influence but not determine the approach of others. Recognition of this reality led to the development of the approach to international cooperation that was discussed in Chapter 2. This chapter takes the international framework as given, and discusses the framework developed for Australian decision-making.

Comparing costs and benefits

I posed the questions: Would the substantial costs of mitigation be exceeded by avoided costs of climate change? What degree of mitigation would lead to the largest net benefits?

The costs of mitigation come early. They are relatively straightforward. They can be measured with standard modelling techniques. The benefits of mitigation come late, and only some of them are amenable to quantitative analysis. I expressed concern that quantification of some of the benefits of mitigation would lead to devaluation of the rest. That concern was justified by the way many commentators focused only on what was measured, and neglected important considerations that were discussed qualitatively. *Eʟ ʙᴊᴏʀɴ ʟᴏᴍʙᴏʀʟ*

I introduced a four-part typology, which is illustrated in Chart 3.1. Type 1 covers costs of climate change that are amenable to reasonably precise measurement using market prices. They are built around average (median) expectations of the science of climate change. Type 2 covers costs of climate change that are of a conventional economic kind but for which there is insufficient data to allow precise measurement through an economic model.

I excluded from measurement any effects beyond the twenty-first century. Excluding all economic effects beyond 2100 did not diminish the costs of mitigation, but removed the majority of benefits.

Type 3 effects occur because outcomes may be much more benign or much more severe than the average suggested by scientific analysis.

Chart 3.1 The four types of mitigation benefits

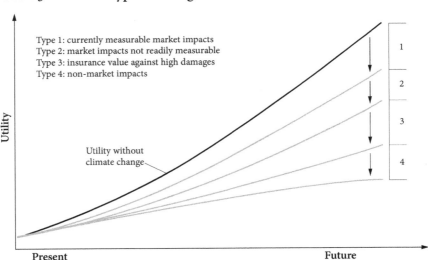

I noted that humans place greater weight on possible catastrophic outcomes than on a similar chance of an exceptionally favourable outcome. Nicholas Stern had handled uncertainty in a different way, by looking at all of the possible outcomes, weighting each by the chance of their occurring, and adding up the weighted possibilities. This 'Monte Carlo' approach to uncertainty undervalued insurance against the worst outcomes. I thought it better to recognise the difficulty of turning concern for severe and catastrophic outcomes into effects on an average, and to ask readers to assess the effects of uncertainty separately from the calculations based on median outcomes. In the qualitative analysis, I expressed the fear that shocks to established political order of the kind and severity that would be associated with unmitigated climate change could cause social and political systems to fall apart.

Type 4 effects recognise that Australians value many things other than their own material standard of living. Examples of such non-market values include environmental amenity, including conservation of species; physical and cultural human heritage; and avoidance of trauma and poverty in other countries. I noted that what matters most is how future generations of Australians, experiencing the full force of climate change, will value these things. Australians in those future times would probably, although not certainly, be materially richer than Australians are today. Therefore, they would likely value increments of material comfort less highly than we do today, relative to environmental amenity, human heritage and the wellbeing of other people.

It still seems right to me to avoid combining effects into a single number. Economists who once focused more strongly on the measurable Type 1 and 2 benefits, including William Nordhaus in his recent book *The Climate Casino*, have come to emphasise more strongly the separate importance of Type 3. Developments in moral philosophy, including the teachings of Pope Francis discussed in Chapter 2, emphasise the separate importance of Type 4 benefits of mitigation.

Introducing the fish

The 2008 Review used a diagram that compared how wellbeing could be expected to change over time if we did nothing about climate change, and if we took action to mitigate its effects.

Human 'utility' or welfare, in which the material standard of living is an important element, rises over time through the usual processes of capital accumulation and increases in productivity that have been proceeding since modern economic growth began a quarter of a millennium ago. The rate of increase in utility slows through the influence of unmitigated climate change. This is the grey line in Chart 3.2.

Mitigation leads initially to a reduction in utility, as it diverts resources from consumption and from investment to increase economic output. So utility with mitigation is initially lower than without. This is the darker line in Chart 3.2. But this is a temporary condition. At some point the cost of mitigation reaches its maximum, and from that point it ceases to hold back growth in material standards of living. There may be a crossover point, after which utility with is greater than utility without mitigation.

The two lines together, with a crossover point, trace out the shape of a fish. Mitigation is justified if the area contained by the tail exceeds the area contained by the body of the fish. The comparison needs to take account of any differences that we attach to the value of a unit of utility today, relative to a unit in future.

A more ambitious mitigation target, or a change that increases the cost of reducing emissions, causes the belly of the fish to fall below where it would otherwise be in the period prior to the crossover point. A more ambitious target also increases utility in later years, through reduced cost of climate change. The shifts in the lighter and darker lines together change the crossover point – in Chart 3.3 they bring it forward in time.

Chart 3.2 The climate change fish – utility with and without mitigation

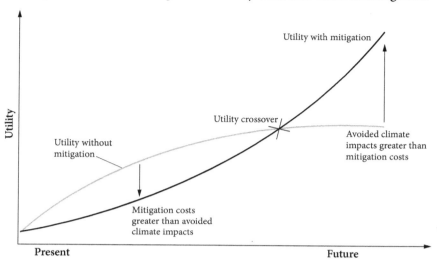

Chart 3.3 The climate change fish – utility with more ambitious mitigation

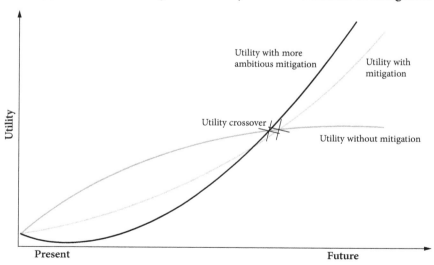

Taking more climate change costs into account – extending the analysis beyond Type 1 and Type 2 effects to take in insurance, heritage and other values – reduces utility in the absence of mitigation. In Chart 3.4, this brings forward the crossover point.

Taking into account costs of climate change beyond the twenty-first century extends the tail. The long lags in realisation of sea-level rise and many other impacts could cause the tail to increase by large amounts.

How have the positions of the fish's back and belly changed with new knowledge since 2008? And has anything happened to change our valuation of the tail relative to the body – to change our valuation of future relative to current welfare?

It turns out that the rate at which future income or consumption or utility is discounted is a powerful influence on the size of the body and tail.

Chart 3.4 Taking more mitigation benefits into account

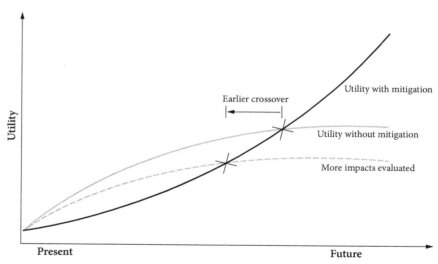

What has changed?

The discount rate

The 2011 Review began with a discussion of the rate at which future benefits are valued relative to present benefits. No subject is duller for most people when they are first exposed to it. No subject is more engaging once its consequences are understood. For those who find it dull, I have separated detailed treatment of the issue into an appendix at the end of this chapter. Those who find it engaging can read the appendix now or later.

There is confusion in the economics literature over the appropriate discount rate – including in the otherwise sound pioneering work of Nordhaus. There is confusion partly because it is complicated. The way in which we value income and utility in the future relative to the present comes into decisions on the costs and benefits of climate change mitigation in several different ways, requiring use of different discount rates.

Three different interest or discount rates are relevant to analysis of the costs and benefits of climate change. The first compares the value of the wellbeing of people living in future with that of people living today. The second is the rate of return to business investment, and measures the income foregone if we divert capital to climate change mitigation that might otherwise have been invested simply to earn future income. The third is the rate at which a carbon price should rise over time, which determines how much of an agreed mitigation effort should be made now and how much left until a bit later.

The 2008 Review found that strong mitigation was justified at discount rates applicable a decade ago.

On the first discount rate, several reasons are sometimes given for valuing a unit of utility for people living now more highly than for people living in future. One is an assertion that humans value things and good exeriences more highly if we can enjoy them now than if we have to wait to enjoy them, and that we should accept that view of value. Nordhaus, among others, has taken that position. It usually leads to an assertion that future benefits should be discounted at the rate of return that people expect to receive from business investments. There is no good reason for valuing utility sacrificed now more highly than the same amount of utility received by people living in future.

Acting consistently with this view of value leads to outcomes that most people would think absurd. I began the 2011 book with an anecdote from the deliberations of the Multi-Party Parliamentary Committee on Climate Change chaired by Prime Minister Julia Gillard: applying business investment discount rates would lead to a decision not to take action even if it was known with certainty that inaction would lead to the early extinction of the species. Economist John Quiggin said in a submission to the 2008 Review that most people value the welfare of their children as highly as their own, and that our children can be expected to feel the same about the welfare of their own children, until the end of human time. That argues against valuing a unit of utility that comes to people in later generations less highly than a unit of utility enjoyed by people living today.

A second and different reason for valuing a unit of utility now more highly than a unit enjoyed by people living in future has validity. In countries experiencing modern economic development, most generations enjoy a higher living standard than their precedessors. It is reasonable to value a unit of income enjoyed by a relatively poor person more highly than the same amount of income enjoyed by a relatively rich person. Presuming continuing growth in living standards generation to generation, a unit of income spent on climate change mitigation now has more value than the same amount of income saved for future generations by avoiding climate change. (This second reason for discounting future utility depends on attaching positive value to fairer income distribution. Consistency requires people motivated by this second reason in taking decisions on climate change policy also to favour redistributive taxation, social security and foreign aid programs.)

The second discount rate – the rate of return on business investment – becomes relevant if some sacrifice of income today leads to lower investment in activities that would earn more income in future. Only part of any current reduction in incomes would lead to lower investment. The economic modelling for the 2008 Review carefully measured the proportion of any change in incomes and expenditure that would take the form of lower investment. It used historical average returns on investment to calculate the loss of future income following increased expenditure on mitigation of climate change.

The third relevant discount rate is the rate of increase in the carbon price. One innovation in the 2008 Review was its reference to the economics literature on optimal rates of depletion of a finite resource. There is a calculable total amount of greenhouse gases that can be released into the atmosphere – the greenhouse-gas 'budget' – that is consistent with agreed goals on limiting temperature increases. So I treated the atmosphere's capacity to absorb emissions as a finite resource, one that could be depleted quickly or slowly, early or late. The resource economics literature said that optimal rates of depletion of the atmosphere's finite capacity to safely absorb greenhouse gases would be achieved if the carbon price rose at the rate of interest.

There has been an historic and large fall in interest rates in the world as a whole and in Australia since the 2008 Review. This stands alongside the fall in the costs of equipment for generating and storing renewable energy as an important change in the economic environment affecting the case for climate change mitigation. The large falls in interest rates reflect structural and not cyclical and temporary changes in the Australian and other economies.

Lower interest rates do not affect the valuation of a unit of utility to people living in future relative to people living now. However, lower interest rates are partly a response to lower levels of productivity growth and investment that have lowered expectation of growth in incomes in Australia and other developed countries. This weakens the case for discounting a unit of utility for people in future because we expect them to be richer than people today. It reinforces the case for strong climate change mitigation.

The historic falls in global interest rates have been associated with a decline in the rate of return on business investment. This has reduced the cost of climate change mitigation – the more so since the main low-emissions technologies use far more capital and have lower recurrent costs than the fossil energy-based technologies that they replace.

The historic falls in interest rates also raise the initial carbon price and lower the rate of increase that minimises costs of meeting climate change objectives. This argues for achieving more of the reduction of emissions early.

So the fall in global interest rates and the structural changes in the economy that are its cause consistently point to lower temperature targets, more emissions reduction and earlier progress towards targets than was justified in the economic environment of 2008. These changes strengthen the case for a 1.5°C objective and for moving early into the emissions reductions that are necessary to achieve that outcome.

Lower interest rates and the structural changes that lie behind them expand the fish's tail relative to the body, and shrink the belly. The size of the tail exceeds that of the body at lower thresholds of temperature increase. They support the case for a 1.5°C objective, and for making an early start on the move to zero net emissions.

Other costs of mitigation

As we move to zero net emissions, ways must be found to replace demand or change supply of all products and activities that generate greenhouse-gas emissions. It may be helpful to remind ourselves of the sectoral composition of the task. We have made substantial progress on land use and electricity emissions. We have gone rapidly backwards on transport and on fugitive emissions from coal and gas production.

A number of other developments, beyond the fall in the costs of capital, have radically reduced expectations of the cost of mitigation in some sectors.

The cost of machines for producing electricity from wind and sun has fallen far more rapidly than anticipated. This has been supported by similarly transformative cost reductions in battery storage of electricity, starting a bit later. These cost reductions are important beyond the electricity sector. They are reducing the cost of decarbonisation through electrification in transport and many areas of industry.

The reduction in the cost of reducing emissions in electricity, transport and industry has had one large incidental effect that should be counted as an additional benefit from mitigation. It has removed what would otherwise have been extreme upward pressure on global prices for fossil fuels following rapid and sustained economic growth in the large developing countries. In the absence of expanded use of renewable energy, such growth in the developing countries could be expected to increase greatly the world price of coal, oil and gas. This would threaten continued economic development in the old style. The expansion of renewable energy will remove this problem. Global coal prices increased about six-fold in response to the huge increase in Chinese demand through the first decade of this century. The moderation of prices for several years from 2012 reflects to a considerable extent the shift of virtually the whole of the subsequent increase in Chinese electricity demand away from fossil energy. It is now possible to contemplate continuation of global economic development over a long period without disruption from limited availability and high prices of energy.

Chart 3.5 Australian emissions by sector

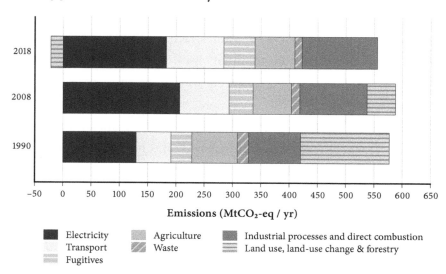

Source: Department of the Environment and Energy, accessed 29 March 2019.

Benefits of reduced climate change

Chapter 2 showed that recent science on the impact of climate change has generally increased certainty without greatly changing average expectations. However, the new work has increased awareness of the importance of some threats – for example, of large reductions in land-based ice in Greenland and Antarctica. This has led to increased insurance value of strong mitigation. Certainly, the economic literature has come to place more weight on the insurance value of avoiding extreme outcomes. This is apparent in Nordhaus's *The Climate Casino*, in which he favours stronger mitigation objectives largely for their value as insurance against low-probability but catastrophic outcomes.

Of more importance in the public discussion has been the clearer understanding of the importance of the non-economic values affected by climate change, the Type 4 objectives. Here the leading contribution has been Pope Francis's *Laudato si'*. The secular discussion has also come to place greater emphasis on avoiding irreversible damage to the earth's natural heritage. More value is being placed on the conservation of plant and animal species. There is greater recognition now that the costs of unmitigated climate change would fall disproportionately on poor people and developing countries, introducing a stronger social justice element into the case for early, strong mitigation. Since 2008, the defence ministries of many countries, including the United States and Australia, have taken the destabilising effects of climate change more seriously as a defence and security issue.

Type 3 and Type 4 benefits of mitigation weigh much more heavily in the general discussion of the mitigation of climate change now than a decade ago. This has lowered the fish's back, shortened the body and lengthened the tail. It has strengthened the case for the 1.5°C over the 2°C objective.

The rise of policy uncertainty

Where the activities of one business or household impose costs on others, economic analysis argues for a tax on that activity at a rate equal to the costs imposed. The cost imposed on others is an 'externality' of the activity. Taxing the externality has been favoured strongly by economists from both the social democratic and economic libertarian (both 'neoliberal' and 'neo-conservative') parts of the politcal spectrum. The most committed devotees of unrestrained market exchange – Friedrich Hayek and Milton Friedman – for example, saw taxing an environmental externality as dealing with a compelling problem in a way that minimised distortion of the market economy.

Costs of reducing carbon emissions to any desired extent would be lowest for the world as a whole and for each country if there were a single carbon price across all sectors in all countries. Costs would be lower still if there were support for innovation in low-emissions technologies, and regulatory interventions to reduce the effects of information and related failures in dissemination of new technologies. Policy proposals with these rationales were embodied in the 2011 Clean Energy Future legislative package and the later agreement to integrate the Australian with the European Union emissions trading system (EU ETS).

The price that gives the largest excess of benefits over costs of mitigation now and over time can be set in one of two ways. The first is through capping the amount of emissions allowed, issuing permits for emissions up to that limit, and allowing trade in emissions entitlements. The market would reveal the carbon price. The emissions entitlements could be allocated in one large lump, or year by year, or quarter by quarter. Intertemporal efficiency would be secured so long as entitlements were tradeable across time, which would see the price rising over time (see the appendix to this chapter). The entitlements could be allocated centrally by an international organisation or at a national or subnational level. The entitlements would be valuable property, whether purchased or gifted to favoured interests. One international or many national public entities could receive the proceeds of any sale that were not gifted to private entities. The allocation of proceeds of auctions, or of free permits, would have a large effect on the distribution of income among and within countries. To the extent that the value is collected by national governments, it can be used to reduce more distorting taxation, as it was with Australian carbon pricing.

Alternatively, the price that leads to reduction in emissions to the target level could be calculated through economic modelling commissioned by the government. The government would sell permits at that price to any participant in the economy who wishes to emit greenhouse gases.

If there were confidence in the stability of the arrangements, and well-regulated markets and taxation systems, and the modelling perfectly anticipated the economic reality, the carbon prices and the amount of emissions reductions would be identical in the emissions-trading and fixed-price systems.

The Australian carbon-pricing regime had a fixed price of $23 in the first year, and a bit over $24 in the second. It was then to become a floating price through integration into the EU ETS. Australian carbon prices were substantially higher than European ones during the two years of their operation, but lower for years before and after that. If the Australian carbon price had survived and been linked to Europe as originally planned, the price would have fallen initially and then risen to well over 2012–14 levels.

The price of carbon is not a measure of the cost of mitigation. The scarcity value of the permits does not disappear from the economy. If a fixed carbon price is enforced, or if permits are sold by auction, the scarcity value of the permits is collected by government and used to reduce other taxes, or to expand public expenditure, or to reduce the budget deficit. Alternatively, it could be given away to private entities in tax concessions or exemptions or concessional or free allocation of permits, in an attempt to win political or personal favours from the beneficiaries. William Nordhaus has demonstrated that pricing carbon is a relatively efficient form of taxation. The economic costs of collecting some revenue in this way are lower than the costs of income taxation at the rates currently applied in the United States. To the extent that carbon pricing reduces the incidence of more distorting forms of taxation, mitigation is an additional benefit. This was captured in the Clean Energy Future legislation that was in effect from 2012 to 2014.

Regulation and funding as an alternative to carbon pricing

Where broad-based carbon pricing is not possible, emissions can be reduced by regulatory and fiscal intervention sector by sector. This will inevitably cost more. The excess cost will be less the wider the sectoral coverage of the intervention, and the more uniform the implicit cost of carbon across interventions.

After the failure of the Obama administration's carbon-pricing initiative in the Senate following its passage through the House of Representatives in 2012, the US Department of Energy calculated a social cost of carbon to guide regulatory interventions to reduce emissions. The department's more recent calculations have identified prices of A$30–50 – a range that overlaps recent European carbon prices.

International trade in emissions entitlements

International trade in entitlements can lower the cost of mitigation in all participating countries. The reduction will be greatest in countries that in the absence of trade would contribute their fair share to a global effort at exceptionally low or exceptionally high cost. The 2008 modelling placed Australia in the second category – a country with higher than average mitigation costs, and with a tendency towards large purchases of international permits. Reductions in the cost of reducing emissions in Australia, in excess of reductions in other countries, now make it likely that Australia would be a net exporter of entitlements if it entered trade with the European Union.

Free trade in permits with the EU would have begun from July 2014 if the carbon price had been retained. The Australian carbon price would now be the same as the European price – higher than the Australian price from 2012 to 2014 (see Chart 3.6 on page 55). Metals processing from renewable energy would expand in Australia and gradually recede in Europe, so long as more or less free trade were maintained.

Australia has an interest in establishing sound foundations for international trade in carbon credits. International trade in entitlements would then be at prices reflecting costs of mitigation all over the world. Trade would establish similar prices in all participating countries. At the 2008 Review, I thought that the Australian interest was primarily in reducing the cost of meeting Australian emissions-reduction targets by importing permits. Now, the interest is more strongly in finding markets for Australian Carbon Credit Units generated by Australian emissions reductions.

Australian business perceptions of the role of international trade in carbon credits has been distorted by the oversupply of credits within the UN's Clean Development Mechanism (CDM). The exceptionally low prices in the CDM emerged from flaws in the system that have been corrected.

Soundly based trade would be conducted between countries with good regulatory systems or accepting good international regulatory oversight. It would be between countries that have emissions-reduction targets that embody fair shares of the global emissions-reduction effort. It would be within rules that prevent double counting of traded entitlements – the emissions reduction cannot count against the commitments of the permit-exporting country.

The trading arrangement between California and Quebec is an example of good practice. That agreed in 2012–13 between the European Union and Australia would have been good practice on a larger scale.

Soundly based international trade can bring benefits to the participating countries and the world as a whole even if it covers only a few countries. It is more productive if trade occurs among countries with emissions trading schemes, but it can still bring benefits if conducted among countries that are achieving their emissions-reduction goals through other means.

Trade-exposed emissions-intensive industries

There is a valid case on economic grounds for compensating trade-exposed and emissions-intensive industries for the effects of carbon pricing or regulatory interventions on their costs. The validity is limited to compensation for the difference between revenue from sales and the revenue that would have been realised if the rest of the world had similarly costly interventions. The relevant part of the rest of the world for these purposes is the countries whose cost structures set the world price as marginal producers – today, mostly China in the industries that are important to Australia. Disciplined processes are necessary to determine whether and the extent to which the conditions for validity are met. The Clean Energy Future legislation provided for review of compensatory assistance arrangements by the Productivity Commission, which would have determined whether adjustments were required to the arrangements in place from 2012 to 2014.

Australian policy after Abbott

For the time being, the retreat from market-oriented policies that has followed the repeal of carbon pricing in 2014 has blocked economically and environmentally efficient approaches to reducing emissions. It has opened the way to costly distortion of the market economy. Among the retreat's many damaging effects is an increase in policy uncertainty, which raises the supply price of investment not only in the low-carbon industries of the future, but in established industries as well.

The path back to good policy does not go straight to restoration of comprehensive carbon pricing. For now, the first step along the path is realistic recognition of the emissions-reduction task. As discussed in Chapter 2, that is zero net emissions for the world as a whole by the middle of the century and in Australia and other developed countries before that.

A second step is recognition that emissions should be reduced most quickly where costs are lowest. For Australia, the low-cost opportunities include the electricity and land sectors.

Carbon pricing is not an economically efficient incentive for investment in the reduction of emissions unless it is expected to endure over the life of projects. Pending development of positive expectations on

Chart 3.6 Carbon prices in emissions trading systems

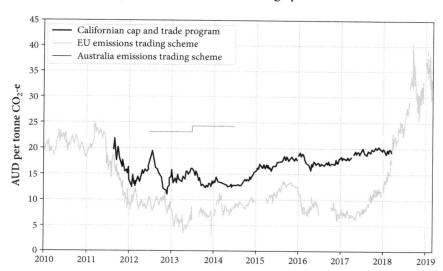

durability of carbon pricing, interventions are more effective if they take the form of prior payments or contracts that do not decline in value if policy changes.

The surviving institutions from the Clean Energy Future package have performed well and now have bipartisan support. The role of the Australian Renewable Energy Agency (ARENA) is to compensate for the benefits from innovation to the society and economy that are not captured by the innovating firm. The role of the Clean Energy Finance Corporation (CEFC) is to lower the cost of capital to activities with large external benefits that would otherwise be denied access to finance because of policy uncertainty. The provision of finance at rates closer to the Commonwealth's cost of capital can also be a means of supporting the low-carbon transition more generally in the absence of carbon pricing. ARENA, CEFC and ERF support, coming through contracts in advance of expenditure, is not vulnerable to subsequent change in policy. Having established their credentials through good performance since their establishment, ARENA, the CEFC and the ERF could have their roles extended more generally to innovation and financing of the post-carbon economy.

There would be gains for Australia and its trading partners from entering arrangements for trading emissions permits with other countries with credible emissions-reduction targets and sound administration of carbon mitigation. This would be assisted by, but does not require, Australia having a domestic trading scheme.

The contemporary fish

So, let us return to the fish in Chart 3.2, and see how the various changes over the past decade have changed its shape and the size of the body relative to the tail.

The belly has lifted by a large amount. There are two big causes: lower interest rates and falling technology costs for renewable energy and transport. Chapter 7 will suggest a third: increased confidence in large-scale land sequestration. Policy incoherence has greatly increased the cost of reducing emissions, but has increased the cost of the old technology as much or more.

The back of the fish has also contracted. The most important cause has been increased awareness of the insurance value of mitigation, and of the value of the natural and civilisational heritage that is threatened by climate change.

Reduced confidence in rising utility from generation to generation has reduced any discount that might be applied to the future relative to the present. This has enlarged the tail and increased its size relative to the body.

We now have a skinny garfish with a humpback's tail rather than a plump Port Phillip Bay snapper.

A 2°C objective was amply justified in 2008. The changes in the shape of the fish justify substantially stronger and earlier mitigation today. The framework used in 2008 would now justify a 1.5°C objective.

It will not be possible to get back to broad-based carbon pricing linked to international markets soon. But move forward we will, with other policies, as the saving grace of community concern demonstrates its tenacity and influence in our own and many other political systems.

Appendix to Chapter 3

Discounting the future

There is considerable confusion about the crucially important question of how we should value 'income' or 'consumption' or 'utility' or 'wellbeing' at one point of time compared with those same desirable things or conditions at another point of time. It is crucially important because the costs of taking action to reduce climate change come early and the benefits come much later – many of them one and more centuries later.

The Review took a clear and distinctive approach to this question which has stood the test of time. In this appendix, I seek to do three things. First, I explain the approach to discounting taken in the two Reviews, and how this differs from approaches used by others. Second, I discuss how interest rates have fallen to historically low levels over the past decade or so, and how the fall affects the discount rates that we should use. Third, I explain how the resulting changes in discount rates affect the decisions we make on climate change mitigation.

Three types of discount rate

Much of the confusion in the public discussion, and too much in the economics literature over the appropriate discount rate – including in the otherwise sound work of Nordhaus (2008) – arises because others have not clearly separated three different kinds of discount rates that are relevant to measuring different aspects of the costs and benefits of taking action to mitigate climate change.

The first relates to differences that are applied to the value of a unit of utility or consumption or income or wellbeing for people in future relative to a unit for people today. I call this the Intertemporal Utility discount rate.

The second relates to the income we can earn by undertaking more income-generating capital expenditure – and that we forgo by undertaking less. I call this the Business discount rate.

The third relates to the rate at which a carbon price needs to rise over time if it is to provide incentives for reducing emissions at the pace that minimises overall costs – what I call the Hotelling discount rate. The Review was a pioneer of application of a budget approach to setting limits on emissions. It calculated a global 'budget' of carbon dioxide–equivalent greenhouse gases – the atmosphere's capacity to absorb emissions without breaching agreed limits on temperature increases as a finite resource. The budget could be depleted at a high rate and quickly and exhausted rapidly. Or it could be used at a low rate, slowly and over a longer period of time. The rate at which the budget was exhausted would be determined by the rate at which the carbon price rose. I call this the Hotelling discount rate, following the pioneering work of US economist Harold Hotelling on optimal depletion of a finite resource. The Hotelling rate is the rate of increase in the price of a finite resource that leads to the most economically efficient rate of depletion. It is a market interest rate. A low Hotelling rate would cause a higher initial carbon price, more early reductions of greenhouse gases, and mitigation continuing over a longer period. A high Hotelling rate would cause a lower initial carbon price, less early reductions of greenhouse-gas emissions, and an earlier movement to zero net emissions.

The Garnaut–Treasury general equilibrium modelling of the Australian economy applied in the 2008 Review built on the Centre of Policy Studies model developed by Peter Dixon, Phillip Adams and colleagues then at Monash University, now at Victoria. The model showed that mitigation would cause Australian expenditure to fall moderately below levels that would otherwise prevail in the early years of effort. Part of the fall would be reflected in lower consumption, and part in lower investment. To the extent that it was reflected in lower investment, future national income would be lower, since the investment forgone would not earn a return. The amount of forgone income per unit of forgone investment was defined by the Business rate of return. This was set at 4 per cent in real terms – the average experience of Australian investment over a long period.

The modelling imposed a Hotelling rate of return of 4 per cent. The similarity of the Business and Hotelling discount rates was coincidental. This reflected my judgement at the time of the rate at which market participants would cause the value of emissions permits to rise over time if a total quantity were issued that was equal to the amount of greenhouse gases that could be emitted without exceeding agreed limits on temperature increases, and if they were free to use them at any time.

The models with Business and Hotelling discount rates embedded in them were used to calculate the costs and benefits of Australia doing its fair share in a global effort to hold temperature increases to 2°C and 3°C. Making that effort required a sacrifice of income and expenditure in the early years, with an augmentation as a result of reduced costs of climate change in later years. To compare the value of the consumption forgone early with the augmentation later, I used Intertemporal Utility discount rates in the range 1.35–2.65 per cent.

Determining the three discount rates

The setting of the Business discount rate is relatively straightforward. The 4 per cent used in the modelling was the average rate of return on business investment over long periods. This is similar to rates used by most economic modelling of the costs and benefits of climate change mitigation, including that of Nordhaus. The Business discount rate is linked to interest rates in competitive financial markets. It can be expected to fall with the interest rate.

In setting the Hotelling rate, I was seeking to simulate how the markets would operate if there were confidence in property rights in emissions permits and the longevity of a trading system. The concept is relevant whether or not the main policy instrument for reducing emissions is an emissions trading system.

In setting the Hotelling rate, I had in mind the global gold market as an analogue of the carbon permits market. Gold markets are old, deep and credible. Futures prices rise year by year in line with medium-term interest rates – both measured in the same currency, today usually in US dollars. If interest rates rise, the spot price of gold (the price today) tends to fall, and the futures price to rise – and rise even more the further we move into the future. If interest rates fall, today's price of gold rises and future prices fall.

The curve that maps current and future prices of gold is called the gold contango. I thought that the carbon price profile that would minimise mitigation costs would take the form of a carbon contango.

I judged that at least for a considerable time, the financial markets would add a premium for uncertainty to the carbon contango by allowing a margin over the interest rate implicit in the gold contango. In retrospect, I may have set the Hotelling interest rate too high. Be that as it may, a lower Hotelling rate would be more appropriate now than in 2008, because interest rates have fallen. A lower Hotelling rate today would mean a higher starting price for carbon, rising less rapidly over time.

The Intertemporal Utility discount rate – how we value the consumption or utility of people in future relative to people today – has powerful effects on judgements about how much sacrifice it is appropriate for people today to make to improve the lives of people living in future.

A high Intertemporal Utility discount rate, as proposed by some participants in the discussion, effectively rules out any sacrifice of comforts now to buy utility for later generations. If the Intertemporal Utility discount rate is 7.2 per cent, a unit of utility in 2020 has 7.2 per cent less value than a unit in 2019. The simple arithmetic says that we then value a unit of utility in 2029 at half of a unit of utility today. A unit of utility in 2039 will have one quarter of the value of a unit today; and in 2119 less than 1 per cent. We wouldn't spend anything much at all today to avoid a catastrophic outcome for humanity in a hundred years time.

The Intertemporal Utility discount rate has three components. One is the rate of pure time preference (valuing the utility of people living in the future less highly simply because they are living in the future). I see no good reason for applying a positive rate of pure time preference. This is consistent with widely held human values when people think about the consequences of doing anything else. It has strong support in the philosophically sound economics literature. The second is discounting future utility because people living tomorrow can be expected to be richer in material comfort than people living today. It is reasonable to attach greater value to a unit of consumption by relatively poor people (us today if incomes are rising over time) than by relatively rich people (people living in future). The third component reflects the small but regrettably positive probability that humans will not be around to enjoy the future benefits of today's sacrifice of utility – the extinction of our species before people have enjoyed the reduced damage from climate change.

I applied an Intertemporal Utility discount rate for Australia in the range of 1.35 per cent to 2.65 per cent, plus inflation. This comprised the sum of the three parts mentioned in the previous paragraph. I used a zero rate of pure time preference. A small component recognised the possibility of early extinction (0.05 per cent). The biggest component (1.3 to 2.6 per cent) came from expectation that Australians' material living standards would be higher in the future, so that a unit of consumption sacrificed for mitigation now was less valuable than a unit conserved by mitigation for consumption in the future.

The Garnaut–Treasury modelling generated a 1.3 per cent per annum increase in per capita consumption through the remainder of this century. This meant that average material living standards would be very much higher in 2100 than today. I argued that the 1.3 per cent discount rate at the lower end of the range was consistent with community preferences for equitable income distribution as revealed in decisions on progressive income taxation and other redistributive measures. The 2.6 per cent would indicate that Australians placed much higher value on equitable income distribution than was reflected in decisions on progressive income taxation and social security. I included the high end of the range despite its implausibility to demonstrate that the case for 2°C mitigation was strong even on the most unfavourable of credible approaches to discounting future utility.

The Review drew attention to several reasons for caution about discounting future utility on the grounds that people in the future will be much better off in material terms than people today. First, with Australian incomes in 2007 and 2008 high in the China resources boom and with productivity growth much lower in the early twenty-first century, we could not be certain that material consumption per person really would be higher in future. It was less likely that future generations would enjoy higher material living standards than we do today in a world suffering losses from climate change. In 2008, the possibility that material living standards would be lower for future than for contemporary Australians was mentioned for completeness. It seems more relevant today, after half a dozen years of stagnant and sometimes negative real wage growth for ordinary workers.

Second, people are likely to value non-market human and natural heritage services more highly relative to increases in conventional consumption in future if incomes and material consumption are much higher than they are today.

Third, climate change may greatly diminish human and environmental heritage values that are enjoyed by future generations. As a result, we cannot be sure that, even if there were much higher material consumption, the average utility of people in future will be greater than the average utility today.

Fourth, more than any other developed country, Australia, if past and current trends continue, can be expected to have a much higher population at the end of this century than today, as a result of immigration. There is a case for valuing a percentage point increment of consumption at the end of the century substantially more highly than a percentage point increment of consumption today, because it would be enjoyed by more people.

None of these qualifications was taken into the main analysis. If they had been, each would have strengthened the case for strong mitigation. None was required to make the case for Australia playing its full role in a global effort to achieve the 2°C objective.

The Nordhaus criticism of normative rates of return used by Nicholas Stern confuses the Business rate with the Intergenerational Utility discount rate. They are conceptually different, and have different roles.

While the economic literature on carbon pricing before the Garnaut Review generally accepted that the price should rise over time, there is confusion about the analytic basis of the rising price. It is sometimes argued that uncertainty about the appropriate price justifies a low start, so that the losses from a false start are low. Rigorous analysis does not support this argument: the possibility that the costs of climate change may turn out to be much lower or much higher than the mean of current expectations may justify a higher starting price. (Normal risk aversion causes value to be placed on insurance against the worst cases.) It would never justify a lower starting point.

Separately, Nordhaus and others suggested that the optimal price should rise over time because the marginal cost of damage is rising with increasing accumulation of greenhouse gases in the atmosphere. That, too, is an erroneous explanation for a rising price. An early tonne of carbon dioxide emissions does as much damage as a late one. (The analysis is different for some of the short-lived and high-impact green-house gases.)

The carbon price should rise simply because income is forgone for no good reason if the atmosphere's capacity to safely absorb green-house gases is not used at the rate of depletion that minimises costs of mitigation.

The historic reduction in interest rates

Market interest rates have fallen in Australia and the world as a whole over the past decade to the lowest ever. I have explained elsewhere why I think this is a permanent change.

For most of the quarter-millennium of modern economic develop-ment, real interest rates on low-risk securities varied between around 2 to 6 per cent in real terms in the leading capital markets. A presumption that real rates were stuck at around 5 per cent formed the foundations of Thomas Piketty's pessimistic prognosis on the future of capitalism, as outlined in his 2017 book *Capital in the Twenty-First Century*.

The Piketty view contrasts with that of John Maynard Keynes, the greatest and most influential economist of the twentieth century. Keynes expressed the view that under mature capitalism (postulated to lie a century ahead when he wrote in 1931), capital would be cheap and abundant, with real interest rates near zero.

Long-term real interest rates set in capital markets have declined decisively in the last several decades, and emphatically over the past decade. Rates on government and other low-risk debit are near zero or negative in mid-2019 in all substantial developed economies including Australia.

Nominal interest rates have varied with inflation. In times of little or no inflation – outside the global wars and their immediate after-maths, and in the immediate aftermath of the breaking of the Great Inflation that ran from the late 1960s until the early 1980s – long-term nominal interest rates in major developed economies have fluctuated around 3 to 7 per cent per annum. They have declined consistently over the past several decades, to near zero in many developed countries for much of the past decade, and to the lowest levels ever in 2019.

At first, the low real rates since 2008 were attributed to the high unemployment and unconventional monetary policy adopted after the Global Financial Crisis. However, real rates have remained low in the United States despite the return of low unemployment and the phasing out of unconventional monetary policy. This has surprised central banks, business economists and many mainstream commentators on the economy.

Major structural changes in the global economy explain the historic downward shift in interest rates. The low rates deriving from these structural changes will vary over time but will not be reversed on a sustained basis. A continuation of the fiscal profligacy of the current US administration may lead to a loss of confidence in US government securities for a while, but a rise in interest rates in such circumstances is likely to be temporary.

Effects of lower rates on mitigation decisions

The fall in market interest rates lowers the Business and Hotelling but not the Intertemporal Utility discount rates. The main low-emissions industries are much more capital-intensive than their high-emissions alternatives. The fall in interest rates has greatly reduced the cost of building new low-emissions energy, transport and industry systems and of investment to store carbon in the landscape. For example, nearly all of the costs of renewable energy and storage are capital costs, while fuel represents the main cost of coal and gas power generation. Low interest rates therefore not only reduce the cost of low-emissions production, but improve its competitiveness against high-emissions alternatives.

Lower interest rates through their effects on the Business rate shrink the fish's body. They greatly strengthen the case for Australia playing its full part in the global effort to hold increases in temperatures as far as possible below 2°C. Lower interest rates, through their effect on the Hotelling rate, argue for doing more of the required mitigation early. The way that lower interest rates strengthen the competitive position of low-carbon technologies is discussed at greater length in Chapters 4, 5, 6 and 7.

THE ELECTRICITY TRANSFORMATION

The early and orderly movement to zero-emissions electricity is the cornerstone of the decarbonisation of the Australian economy. It is also the foundation for Australia's emergence as a superpower of the post-carbon world economy.

In 2008, electricity generation accounted for nearly 40 per cent of Australian emissions. A decade later, electricity emissions had fallen significantly in absolute levels, to about a third of the total (see Chart 3.5).

The decarbonisation of electricity is much more important than that third of Australian emissions. It also provides a path to near-zero emissions in transport (about a sixth of emissions), much of industry (a quarter) and part of fugitive emissions (a tenth).

Furthermore, the decarbonisation of Australian electricity is important globally, over and above the large contribution it would make to cutting local emissions. Australian renewable energy is a path to low-cost reductions in emissions in the rest of the world, through three mechanisms. First, and most importantly, as we will see in Chapter 5, it will reduce the global cost of decarbonising production of aluminium, iron, steel, and other metals (7 per cent of total global emissions come from converting iron ore into metal). Australia is the largest exporter in the world of mineral ores requiring energy-intensive processing for conversion into metals. Australia in the post-carbon world could become the locus of energy-intensive processing of minerals for use in countries with inferior renewable energy resource endowments. Second, there are opportunities for export of hydrogen produced by electrolysis from renewable energy, through liquefaction or through ammonia as a hydrogen carrier. The natural markets are the renewable-energy-resource-poor countries of Asia, notably Japan and Korea. Third, improvements in technology and reductions in the cost of high-voltage direct-current transmission of electricity are likely eventually to make Australian supply a commercially viable alternative to some high-cost domestic generation of zero-emissions electricity for parts of Indonesia, Singapore and then the mainland of Asia.

After a long climb, electricity emissions fell through the periods of confident expectation (2008–12) and then practice (2012–14) of carbon pricing. In 2018 they were a touch above levels in the last year of carbon pricing. They are lower so far in 2019.

Difficult politics

The electricity transition has been contentious politically.

Electricity became a prime focus of political concern partly because utility price changes are lumpy, sometimes coming only once a year. More importantly, carbon pricing coincided with electricity prices rising for other reasons.

There were four main reasons why electricity rose far more rapidly than other prices after current market arrangements were established in 2006. First, major regulatory mistakes following the privatisation and corporatisation of electricity utilities greatly increased the cost of electricity transmission and distribution to users. Second, making Queensland and New South Wales coal (once dedicated to use by state electricity utilities) and eastern Australian gas exportable raised their

Chart 4.1 Electricity emissions 1990 to 2019

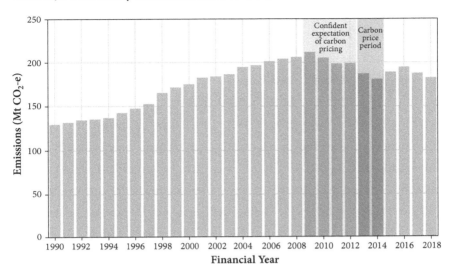

prices to Australian power generators by hundreds of per cent. This influence on prices was even more powerful than it would have been because the apogee of the old, investment-intensive model of Chinese economic growth in the decade from 2003 raised global demand for fossil energy prodigiously, and therefore international gas and coal prices. Third, the domination of each state market by a small number of generator-retailers led to oligopolistic pricing, rather than the competitive market that had been presumed in setting up the system.[1]

As I discussed in my 2013 book *Dog Days*, the China resources boom also raised to historically unprecedented levels the Australian real exchange rate. Together with the increase in energy prices, this damaged the competitiveness of energy-intensive export and import-competing industries beyond the effects of higher electricity and gas costs.

Companies in these industries were compensated for cost increases from carbon pricing by other elements of the Clean Energy Future package, in a way that was absent when gas and electricity prices rose even more in the years following the removal of carbon pricing in 2014. This reality was lost in the fog of domestic media obfuscation and among the foghorns of political confrontation.

Finally, uncertainty about policy through the long political contest reduced investment in electricity generation, reduced supply and increased wholesale prices. It delayed and distorted what would otherwise have been routine regulatory development to support power security and reliability as we moved from synchronous coal and gas generators to increasing proportions of intermittent renewable energy.

From 2009 to 2014, policy support led to considerable progress in reducing electricity emissions, despite the policy disputation and uncertainty.

In the future, the transition to low-carbon electricity will be driven by reductions in the cost of renewable energy and storage. These have two sources. One is the fall in the rate of return required to justify investment, as discussed in Chapter 3. The other is the reduction in equipment costs for renewable energy and battery storage. These economic factors would realise some of their potential in a stable policy environment. They would be truly transformative for Australian decarbonisation and prosperity if the policy environment were economically efficient as well as stable.

In this chapter, I discuss the challenges of the energy transition as they appeared in 2008 and 2011. I then look at how the combination of policy incentives and the increasing competitiveness of renewable energy has supported some decarbonisation over the past decade. I examine the challenge that wind and solar power intermittency poses for power security and reliability. And I conclude by looking forward to the policies for electricity that could support realisation of Australia's immense opportunity to do well economically in the zero-emissions world economy of the future.

The energy transition in 2008 and 2011

In 2008 I noted that Australia's resource base placed it well for the energy transition: it had a wide range of high-quality renewable energy resources and economically favourable opportunities for geosequestration of emissions from traditional coal and gas generation. With a 2°C temperature objective, zero-emissions power would supply more than half Australia's requirements by 2030. Wind and solar would account for most of the increase. Gas would be important in balancing intermittent solar and wind energy. Australia's hydro-electric resources and potential for pumped hydro-electric storage (PHS) in the Snowy Mountains and Tasmania, and perhaps its proximity to the immense hydro-electricity resources on the island of New Guinea, would play big roles in balancing solar and wind. Gas and coal with carbon capture and storage (CCS) would be important before 2030, and expand rapidly after that.

The 2008 Review also calculated that Australia was not an economically logical early home for nuclear power generation. Australia's rich renewable energy resources, and the transport economics that argued for the use at home of energy resources for which international transport costs were highest, made other sources of zero-emissions electricity more economically efficient for a considerable while. Nuclear energy would enter the picture from the 2030s if costs determined its role.

The Review criticised and suggested reform of a flawed regulatory system that had increased costs of electricity transmission and distribution. A new approach to investment in long-distance high-voltage transmission was necessary for securing supply from low-cost renewables and for security and reliability. Public-sector planning was required to get the balance right between improved interconnection and minimising costs to power users. Good outcomes required scale-efficient network expansion rules which removed disincentives to pioneering investment in regions with high-quality renewables resources that were not served by established transmission.

Wholesale electricity prices were expected to rise as a result of carbon pricing. The anticipated price increases would be exacerbated at the retail level by the distortions in network and monopolistic retail pricing. Australia's international competitiveness in energy-intensive industries would decline for a while, as unused potential for low-cost renewable power generation in developing countries was utilised. But the quality of Australia's renewable energy resources would reduce its costs relative to other countries over time. After the standard, low-cost renewable energy resources in developing countries (for example, hydro-electricity in Congo and Papua New Guinea) had been exhausted, Australia would re-emerge as a competitive location for investment in energy-intensive industries processing Australian resources.

Some large uncertainties were highlighted. The future of coal for domestic power generation and export depended on the commercial success of CCS. Without it, the role of coal would decline rapidly after 2030. This would require support for alternative livelihoods in coal regions, particularly the Latrobe Valley.

Reductions in the cost of capital goods for solar and wind were built into the modelling. After consultation with experts in Australia, Europe, the United States and China, we estimated reductions of a few per cent per annum for photovoltaic electricity systems, but noted that technological improvement could increase this. Higher electricity prices would slow the increase in electricity demand. Later, there would be much faster growth in demand as electrification became the low-cost path to decarbonisation of transport, much of industry and household heating.

Some parts of the electricity story had already started to change by the time of the review update in 2011. The problem of wasteful overinvestment and excessive pricing for network infrastructure was more acute than in 2008. So was the cost to consumers of monopoly in retail markets. I noted that in the political environment of the time, the resulting increases in electricity prices were likely to be blamed on carbon pricing.

The 2011 Review set out many advantages of deepening and broadening long-distance transmission.[2] This would increase effective retail competition and reduce the pricing power that oligopolistic suppliers had in some regions. Geographic diversity would help to balance intermittent wind and solar. In some locations, demand peaks related to extreme weather could be served by surplus capacity in other areas. Greater sharing of generation reserves would help to meet diverse demand peaks and to insure against generator failure. Storage in one region – including in the hydro-electricity systems in the Snowy Mountains and Tasmania – could be made available to a wider market. Wider transmission would be able to manage more easily the early retirement of coal generation without shocks in regional markets.

The call for new investment in long-distance high-voltage transmission seemed to contradict the criticism of wasteful overinvestment in the network. I explained that it did not. What was needed was more wisely allocated investment: substantially less network investment overall, but more in long-distance high-voltage interconnection. The cost of new transmission could be reduced by allowing roles for investors who were prepared to accept lower returns on capital expenditure.

Gas generation, including for peaking, was still expected to expand in 2011. However, the establishment of an east-coast export industry would now push up gas and therefore electricity prices. Increases in the cost of nuclear plants had diminished the likely role of nuclear in Australian electricity generation.

So how has the outlook changed since 2011? One big shift is reduced expectation that CCS for coal and gas can play a large role in the

Australian electricity transition. It was clear in 2011 that the future of coalmining in Australia depended on successful development of low-cost CCS. Australian coalminers attached much more importance to blocking action to reduce global warming than to building a long-term future for their industry. Despite the commitment of large financial support from the Australian government, the Australian coalmining industry hardly invested at all in the research, development and commercialisation of CCS.

It now seems that CCS will still have a role, but not in capturing and storing emissions from coal-fired power generation. It is likely to be important in securing negative emissions where carbon dioxide is captured from bioenergy combustion – known as bioenergy carbon capture and storage (BECCS). CCS may have some role with coal power generation in favourable locations, where exceptionally low-cost fossil energy is adjacent to excellent geosequestration sites. It will play a larger role in gas generation in favourable locations. And it will have some role in capturing emissions from coal- and gas-based industrial activities.

Spurs to decarbonisation
Policy
The renewable energy target (RET) was expected to provide the main early impetus to the electricity transformation. The RET was originally legislated by the Howard government. It was greatly strengthened in 2009 as a requirement to produce 41 terawatt hours of electricity per annum of renewable energy by 2020 – expected at the time to be 20 per cent of power supply through the grid. The Clean Energy Future package, legislated in late 2011 and to operate fully from 1 July 2012, implemented the main recommendations of the 2011 Review on electricity. A rising carbon price and a static renewable energy target were expected to make the latter redundant from the mid-2020s.

The initial fixed price of $23 was consistent with the price that the 2008 modelling had suggested would meet a 3°C temperature objective. This compared with about $40 required for 2°C.

The carbon price would reduce the value of renewable energy certificates under the RET to zero by the mid-2020s. The carbon price was more economically and environmentally efficient than the RET that it would replace. Whereas the RET favoured renewable energy indiscriminately over all other forms of energy, carbon pricing would favour less emissions-intensive generation at every margin: black coal over brown; gas over black coal; less-polluting over more carbon-intensive thermal power plants; CCS in proportion to its removal of carbon emissions; renewables over the least-polluting thermal generation. It would therefore meet any emissions target at lower cost than the RET, which did not discriminate among more- and less-polluting coal and gas generation.

The RET was mainly directed at large-scale generators supplying power through the grid. A small-scale component (the Small-Scale Renewable Energy Scheme, or SRES) supported small systems up to 100 kilowatt hours. This complemented state subsidies for rooftop solar that were mostly removed from new investments after the introduction of carbon pricing. The SRES was initially expected to contribute 4 terawatt hours of annual electricity supply but was not capped at that level.

The Australian Renewable Energy Agency (ARENA) was to provide grants to expand private investment in innovation. The Clean Energy Finance Corporation (CEFC) was to increase availability of debt for investment in low-emissions energy. Several Commonwealth initiatives were directed at increasing energy efficiency.

It was expected that decarbonisation would proceed substantially more rapidly in electricity than in other sectors, simply because the cost of reducing emissions was lower.

The election of the Abbott government in 2013 opened a period of extreme policy uncertainty. Carbon pricing ceased in July 2014, but other elements of the Clean Energy Future package continued in some form. The Abbott government appointed a committee to review the RET, with a view to its abolition. The committee was chaired by a businessman, Dick Warburton, who was known for strong personal beliefs that scientific atmospheric physics was flawed. The committee commissioned modelling of the effects of the RET. To its surprise, the modelling showed that abolition or truncation of the RET would lead to higher electricity prices. This was consistent with other reputable modelling of the effects of increased renewable-energy supply on power prices. The reason for this result is that solar and wind generators' costs are no higher if they are generating power than if they are not. They therefore bid into the wholesale market at a zero price (or less where there are revenues from the RET or costs of turning off generation). As a result, expanded supply from solar and wind could only reduce and never increase market prices in the short term.

The long-term dynamics are more complex. Increased wind and solar generation lowers the price of power when it supplies the market in large quantities and demand is low. It reduces the average wholesale price of power and therefore the profitability of coal generators, which cannot easily vary by large amounts their outputs in response to fluctuating prices. There are pressures for closure of the least profitable coal generators. The closure of each coal plant is then followed by a lift in market prices, until the increase is eroded by more wind and solar and the cycle continues.

Following the Warburton review of the RET, the government succeeded in reducing the target from 41 terawatt hours to 33 terawatt hours, with the support of the Labor Opposition in the Senate. A period of extreme uncertainty while these matters were settled led to a hiatus in investment, and for a while doubts about whether even the much lower renewable target could be achieved by 2020. Once the new target was legislated, investment in renewables rose to unprecedented levels, demonstrating that the 41 terawatt hours target would have been comfortably met by 2020 had it been left in place.

ARENA played an important role in supporting groundbreaking Australian research and development. It also accelerated Australian applications of technologies that were established abroad, including grid-scale solar photovoltaic generation, pumped hydro storage and battery storage.

The CEFC helped to reduce the effects of the inhibition that policy uncertainty would otherwise have imposed on lending to renewable energy projects. It provided debt for longer periods than would otherwise have been available. It increased confidence of other, especially international, lenders. It operated profitably.

State and territory governments had agreed to dismantle most of their own emissions-reduction programs as part of the Clean Energy Future package. The repeal of carbon pricing brought state and territory governments back into the field, with the ACT, Queensland and Victorian governments introducing major programs for purchasing power from new renewable energy projects.

The combined effect of a truncated but effective RET, the activities of ARENA and the CEFC, and the power purchase schemes of state and territory governments drove rapid expansion of renewable energy (see Chart 4.2).

In its two years of operation, carbon pricing encouraged less emissions-intensive power at every margin, just as it was designed to do. It took some of the weight off the RET in those two years, reducing renewable energy certificate prices.

Rooftop solar in households and small businesses grew rapidly from about 2010 (see Chart 4.2). It continues to do so, supported by new programs in some states after the repeal of carbon pricing. It was initially driven by the combination of superior solar resources (more strongly in Western Australia, South Australia and Queensland than in

Chart 4.2 Australian renewable power supply 2008–18

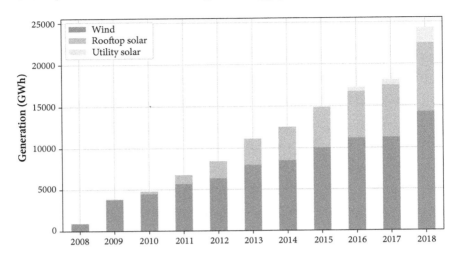

Chart 4.3 Total costs of solar PV per megawatt hour (real A$2019) as perceived in 2011

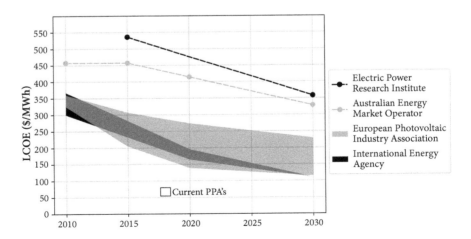

the southern eastern states), strong community interest in the energy transition, official support from the Commonwealth's SRES, and state feed-in tariffs while they were in place. These forces were magnified by falling costs for photovoltaic panels and installation, and rising costs of power through the grid. Output from rooftop solar systems is headed towards greatly exceeding the 4 terawatt hours by 2030 anticipated when the SRES was introduced. By 2018, there had been 8 gigawatts of rooftop solar PV installed in 2 million premises (mostly houses), with 1.8 giga-watts in 200,000 premises in 2018.[3] A higher proportion of households is utilising rooftop solar in Australia than in any other country.

Changing economics

The 2008 and 2011 Reviews anticipated that the total costs of power from solar and wind would be higher than the operating costs of fossil-fuel generators. But even on those expectations, an expansion of renewable energy could lower wholesale prices. The apparent contra-diction between higher overall costs and contributions to lower prices is resolved through understanding price determination in the electric-ity market. Renewables supply reduces prices to near zero whenever it is the source of marginal output to clear the market, as it has been from time to time in South Australia from 2016 and in all mainland states in 2019. It also contributes to lower prices at some other times. Investors in renewable energy recoup their costs from times when prices are higher because sources of energy with higher operating costs (espe-cially gas) are setting market prices. Until the RET requirements have been met, they also receive income from the RET.

Free lunches are rare in real economies. Lower electricity prices from increased proportions of renewables driven by the RET are not a free lunch. They are paid for by the established generators receiving lower power revenues and profits as a result of the renewables contri-bution to lower power prices.

Chart 4.4 Total costs of wind per megawatt hour (real A$2019) as perceived in 2011

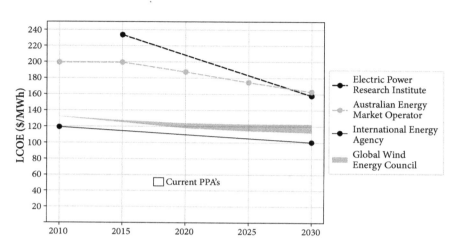

Chart 4.5 AEMO's persistent overestimation of future capital costs of solar PV (real A$2019)

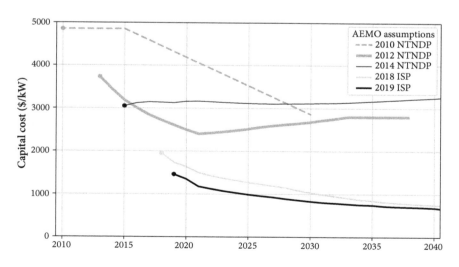

By about 2018, it was clear that the underlying economic realities had changed. In favourable circumstances – where the investor had access to capital at internationally competitive costs, and where projects utilised high-quality resources with access to competitively priced transmission – renewable generators were producing power at much lower total cost than the gas alone for conventional generation. In these same favourable circumstances, renewable generators were producing power at total costs comparable with those of coal generators' operating costs alone, in all states except Victoria. Victoria was different, because lignite from the Latrobe Valley was not exportable and therefore available cheaply for local use.

The large-scale use of intermittent renewable energy required augmentation by flexible sources of power at other times to 'firm' the renewable energy. Firming could come from several sources: geographic diversification of renewable energy sources; gas peaking; pumped hydro and battery storage; and demand management. The cost of 'firm' power was higher than that from solar or wind when the renewable generators were operating.

By 2019, the total costs of coal generation were so much higher than the total costs of solar and wind plus firming that there was no prospect of commercial investors building new coal generators without large direct subsidies – in addition to the subsidy inherent in the failure to tax external environmental costs.

Charts 4.3 to 4.6 demonstrate the dramatic changes in expectations of renewable energy costs between 2011 and today.

Chart 4.3 shows estimates of future total costs (capital plus recurrent) of solar PV electricity by credible agencies at the time of the 2011 Review. Actual prices in recent power purchase agreements (PPAs) in Australia are shown as the grey square – far below the lowest of the estimates for 2019 at the time of the 2011 Review. Chart 4.4 provides a similar perspective on wind.

Chart 4.6 AEMO's persistent overestimation of future capital costs of wind (real A$2019)

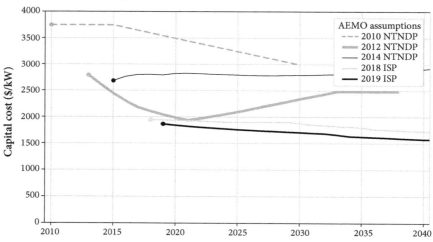

Chart 4.7 The learning curve for solar PV

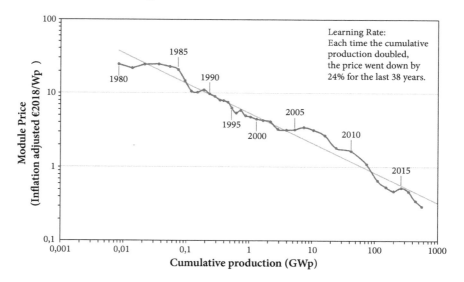

Official expectations of future capital costs of solar and wind chased the realities down but remained a long way behind. (Charts 4.5 and 4.6). The curves setting out AEMO's expectations of future costs fell each year, but never anywhere near as rapidly as actual costs.

The unexpectedly rapid fall in costs of solar and wind power had two causes. One was the fall in cost of funds for investment, as discussed in Chapter 3. The other was the decline in the cost of equipment. Chart 4.7 shows the dramatic and continuing fall in the cost of solar PV equipment with increases in the scale of production, the shift in the locus of production to China, and continuing improvements in manufacturing techniques.

The rapid fall in solar PV and less dramatically in wind and battery storage costs is a triumph for climate policy and the modern global economy. European climate change policies, with Germany leading the way, led to rapidly growing demand for PV panels in the early years of the twenty-first century. Europe accounted for an overwhelming majority of the cumulative global demand for the first decade of the century. It had the largest accumulated installations until 2016, when it was overtaken by China.

European demand flowing from climate policies and open trade policies created an opportunity for new manufacturers all over the world. The Australian tertiary education sector had opened access to international students in 1986, and large numbers of Chinese students benefited from Australian education in the new energy technologies. Graduates in electrical engineering (most notably from the University of New South Wales) returned to China and played leading roles in new solar PV manufacturing companies that quickly became suppliers to Europe, Australia and elsewhere. Costs came down with the scale economies and learning that accompanied expanded production. From 2013, policy support and falling costs contributed to rapid expansion of solar installation in China itself. Within a few years, China was the home of

Chart 4.8 Total solar PV costs versus costs of coal fuel alone

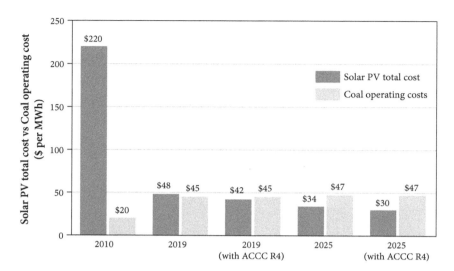

the largest accumulation of installations. It was by then by far the largest source of PV panels. In China itself, a virtuous circle of expanding output, falling costs and expanding demand was well established.

Australian universities, and more generally Australian education and research, benefited from the role they played in the global success of solar PV. Australia also benefited from rapidly falling costs of solar PV.

There are similar stories of the development and transfer of knowledge, global cooperation, and falling costs through increasing scale of production in wind power and battery storage.

The competitiveness of renewable energy in Australia was enhanced by the big increases in coal and gas prices to export parity levels. The increase in coal and gas prices through stronger links to the international market began early in the century and for coal is not yet complete.

The lift in eastern Australian domestic gas prices from the lowest in the developed world to export parity (and for a while to higher lev-els), and to much higher levels than in the United States and other gas-exporting countries, came suddenly with the ramp-up in exports of liquefied natural gas from Gladstone, Queensland, from 2016. This reduced the competitiveness of the manufacturing industry built on low-cost domestic gas. It also dramatically increased Australian electricity prices at those times of the day when gas generation was necessary to meet demand. The Australian Competition and Consumer Commission (ACCC)'s 2018 report on electricity pricing estimated that the increase in gas prices raised gas electricity generation costs by $11 per megawatt hour for every $1 increase in domestic gas prices. Average gas prices more than doubled, to over $8 per gigajoule. The increase in gas prices fundamentally reduced the role gas could play in balancing growing proportions of intermittent renewables.

Chart 4.8 also demonstrates that costs would fall further, to lev-els that would make energy-intensive industry globally competitive through the implementation of the ACCC's Recommendation 4. This recommendation addresses the critical question about Australia's capacity to supply energy-intensive industry of the future with globally competitive power: will Australian suppliers of electricity have access to globally competitive capital? Australia's main competitors in inter-national markets for metals and energy-intensive goods more generally usually have electricity systems built with access to sovereign balance sheets. The impediments include oligopoly in debt and energy markets. The ACCC's Recommendation 4 would support only power supply that was available on a firm basis to users, so that renewable energy would need to be accompanied by means of balancing its intermittency.

The ACCC's proposals, of which Recommendation 4 formed part, were a response to the observation that among twenty-nine OECD countries, Australia had gone from having the fourth-cheapest electricity in 2004 to the fourth-most expensive in 2018. Along the way to reversing the decline, it had to address two issues: oligopoly in energy markets, and impediments to globally competitive access to capital.[4]

The ACCC's proposal was for the Commonwealth to underwrite firm power supply contracts at a low price ($45 to $50 per megawatt hour in 2018). Firm power involved meeting all of the customer's requirements whenever needed. The quantity would cover the debt and not the equity component of the new investment. Chart 4.8 interprets that as 80 per cent of capital expenditure in new investment. The underwriting would only be available to suppliers that did not have a strong position (less then 10 per cent) in the markets that new investment would serve, and therefore would increase competition. It would be available to new retailers which had contracts with at least three commercial and industrial customers. The underwriting would cover the sixth to the fifteenth year of power supply commitments. The Commonwealth would administer the scheme through an appropriate agency, perhaps the CEFC. The scheme would be available to all power retailers that met the transparent criteria, and be open for four years.

To have maximum impact on the competitiveness of the Australian industry, if administered through the CEFC, it would need to be operated separately from the general CEFC portfolio so that the Commonwealth's credit standing was reflected fully in the cost of debt.

The ACCC's Recommendation 4 was to be available to support any investment in power generation and storage. It has been criticised for favouring investment in new and refurbishing old thermal power stations. Yes, it could in principle support investment in coal generators. But the contemporary economics of power supply make it unlikely that such investment would be commercially attractive unless government subsidy went well beyond the underwriting of power prices.

In February 2019, prior to the federal election, minister for energy Angus Taylor announced that a number of generation and storage projects were candidates for Commonwealth underwriting. This was a variation on the ACCC theme. Full implementation of the ACCC Recommendation 4 would have more benefits for reducing prices and increasing reliability of power, and increasing the competitiveness of Australian industry. Recommendation 4 could be crucial in making Australian new energy supply not only cheaper than old energy, but also globally competitive.

The solar power costs presented in Chart 4.8 do not include the considerable costs of balancing intermittency. Firming costs for solar power depend on local access to low-cost wind generation, transmission, storage, gas peaking and demand management opportunities. Today, the cost of firming would typically add $5 to $20 per megawatt hour to total costs. For a large industrial operation, firm power supply will require investment in balancing assets, as well as in renewables generation. The availability of Recommendation 4 support would reduce total solar generation costs by about one-eighth. It could be expected to reduce firming costs by a similar proportion.

Global competitiveness in power supply today would require a delivered price in the range of $45 to $55. That will fall in the future, but not as rapidly as the expected decline in Australian renewable energy costs. In the absence of the ACCC's Recommendation 4, globally competitive power supply is feasible now in the most favourable locations. With Recommendation 4, it is feasible now in most major industrial locations outside the capital cities – including Collie–Bunbury in Western Australia, the Upper Spencer Gulf in South Australia, the Latrobe Valley and Portland in Victoria, Newcastle and Port Kembla in New South Wales, and Gladstone and Townsville in Queensland.

The scene has been set for decarbonisation and lower costs for Australian wholesale power driven by the cost advantages of renewable energy. But a trainwreck of regulatory failure has to be cleared before the underlying economics can transform Australian prospects.

The trilemma trainwreck: emissions, prices, security

Reaction against high prices became tangled in the febrile debate over climate change policy after Abbott became leader of the Coalition in 2009.

Chart 4.9 shows large increases in all components of the retail price structure. Network charges continued to contribute the largest proportion of the increase. Wholesale price increases were modest only in comparison with others. Payments for state and federal environmental policies accounted for 15 per cent of the increase from 2008/09 to 2018/19 – an overstatement to the extent that expanded renewable energy supplies reduced the wholesale price of power, without this reduction in prices being reflected in Chart 4.9.

Carbon pricing from 2012 to 2014 only affected the wholesale price component of costs to power users. It did not affect at all the 'Retail' and 'Regulated network' components of the price. It is remarkable that, after a brief dip, wholesale prices have been higher since the abolition of carbon pricing in 2014, even after removing the effects of general inflation. The substantially higher wholesale prices after the repeal of carbon pricing can be attributed to policy uncertainty inhibiting investment in generation and transmission, and the increased gas prices and reduced gas availability that accompanied exports from Gladstone. Policy uncertainty and weakness and the gas problem increased the costs of adjustment to coal power station closures.

From September 2016, public concerns about power prices were joined by security anxieties. There was a statewide blackout in South Australia on 28 September. An extreme weather event – unprecedented in recorded history for its disruptive impact in South Australia – smashed twenty-three electricity transmission towers. This included destruction of towers on three of the four lines linking Adelaide to the power transmission hub near Port Augusta in the north. The state was hit by at least 80,000 lightning strikes, some of which damaged power

Chart 4.9 Retail electricity price 2009/10 to 2018/19 (national summary, real A$2018)

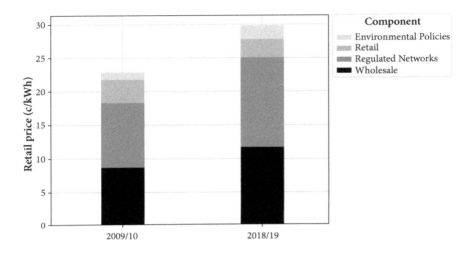

generators. Safety controls automatically shut down a majority of the wind turbines in the state, other generators and the main interconnector with Victoria. The sudden loss of voltage in the system caused a complete shutdown of power supply through the grid.

Australian consumers had paid high network prices for many years for backup ('black start') capacity, to bring the power system back to life quickly if it failed. But these 'black start' systems didn't work when needed most. South Australian consumers had paid millions of dollars year after year to keep diesel generators in place at the western end of the network in Port Lincoln as insurance against failure of the transmission lines. The diesel generators didn't start when they were needed.

South Australians lost access to electricity for several hours – excepting only the small number who had battery systems designed to operate independently of the grid, and the residents of a few places, including Kangaroo Island, which had independent generators that worked when needed. The residents of the Eyre Peninsula, including the fishers of Port Lincoln with their perishable catches, were without power for several days.

Two subsequent episodes consolidated energy security as a priority for contemporary Australia. In December 2016, failure on the high-voltage transmission line from the lignite generators of the Latrobe Valley to Portland blacked out a large part of western Victoria for long enough to cause serious damage to the aluminium smelter. The loss of voltage at the Heywood substation, north of Portland, tripped the interconnector with South Australia, and many South Australian households lost access to power again. And then a heatwave of rare intensity and dimension covered eastern Australia for several days in February 2017, increasing use of electricity and damaging power supply from some thermal generators. This time, New South Wales was at the epicentre of the problem. The largest power user, the Tomago aluminium smelter, went close to catastrophic failure.

The anxieties over energy security from the spring and summer of 2016/17 shaped the initial political and media response to the closure in autumn 2017 of one of the largest and oldest, and distinctively the dirtiest, Australian generators: the Hazelwood lignite-fuelled power station in Victoria's Latrobe Valley. The market, guided by AEMO, managed adjustment to the closure without disruption. However, the withdrawal of much supply capacity compounded the upward pressure on prices that was already present from other causes.

South Australia is the state with by far the highest proportion of intermittent renewable energy. It had a state renewable energy target of 50 per cent by 2025 under the Labor government that lost office in 2018. It exceeded that proportion in 2018. Expanded renewable energy supply has shifted South Australia's position in the National Energy Market from net importer to net exporter. For the most part, the high penetration of renewables in South Australia results from national policies having a more powerful impact there because solar and wind resources are of higher quality than in New South Wales and Victoria. In addition, South Australian coal resources are poorer, leading to higher electricity prices from traditional sources before solar and wind were important.

The blackout led to a political storm over climate and energy policy. Prominent figures in the federal Coalition parties and One Nation blamed the blackouts on the South Australian state government's support for renewable energy. Premier Jay Weatherill decided to move forward with the transition, rather than to retreat. His government set out to make a system with a high proportion of renewables secure and reliable. The policy response included support for what would become the world's largest grid-level battery, and establishment of a 270 megawatt government-owned gas turbine generator as backup to the system. The SA Coalition government led by Premier Steven Marshall formed in early 2018 has maintained a positive approach to the energy transition, and new policy includes fiscal support for small-scale batteries and for an interconnector to New South Wales.

The Council of Australian Governments (COAG) Ministerial Council for Energy is chaired by the Commonwealth Minister for Environment and Energy (at the time Josh Frydenberg), with all state and territory ministers for energy as members. Immediately after the SA blackout, COAG appointed an experienced and able group led by Chief Scientist Alan Finkel to make recommendations on the security and reliability of the Australian energy system as the energy mix changes to reduce emissions.

The most important consequence of the South Australian blackout and the Finkel Review was to elevate security among electricity policy objectives. Generators were henceforth to give three years' notice of closure. There was discussion of markets for demand management and capacity. More attention was paid to safety triggers in wind generators in response to changes in voltage. Prime Minister Malcolm Turnbull elevated discussion of pumped hydro storage as a means of providing security and reliability in the new system. There was increased recognition of the contribution that grid-level batteries and synchronous condensers could play in maintaining frequency. The Australian Energy Market Commission (AEMC) approved a rule change to fix wholesale prices at five-minute rather than half-hour intervals. This would facilitate the use of batteries for grid stability and reliability when it belatedly took effect from 2021. AEMO extended markets for grid stability services. For the first time, there were prospects for the regulatory arrangements facilitating rather than resisting the many changes that were necessary for security and reliability in a low-carbon electricity system.

Security and reliability had always been responsibilities of the regulators reporting to COAG – the AEMC, the Australian Energy Regulator (AER) and the Australian Energy Market Operator (AEMO). The 2016 blackout in South Australia and the subsequent brownouts in Victoria, New South Wales and South Australia marked a failure of that responsibility. The heightened focus on security and reliability from 2017 ended the neglect. AEMO, under a new CEO, Audrey Zibelman, played a leading role in bringing regulatory approaches into line with a future in which intermittent renewable energy would play a central role.

Since the summer of 2016–17, the Neon-Tesla big battery, other batteries, the government's gas turbines and more attentive regulatory agencies have made South Australia possibly the most secure region within the National Energy Market. Reliability has now come into stronger focus: ensuring that there is enough reserve generation capacity for supply to match demand through increasing fluctuations in both.

Reliability, transmission and storage

In 2011, the AEMC rejected the Review's proposal for scale-efficient network extension. The grounds for rejection was that it would increase network costs to consumers. This commendable concern for consumers contrasted with the attitude to conventional investment in the network. Such investment had led to an increase in the Regulated Asset Base by over $40 billion since 2006, a period which saw no increase in demand through the grid. The Grattan Institute later estimated that $20 billion of this expenditure could be categorised as wasteful overinvestment.

There has been little expansion of the interstate transmission capacity in the past decade. This greatly increases the difficulty and cost of maintaining reliability and minimising power supply costs through the transition to low emissions.

Since the blackouts and brownouts of 2016–17, there has been belated recognition of the importance of new long-distance transmission. The Reviews proposed two initiatives to minimise the cost of expanding transmission infrastructure in the new circumstances, beyond the scale-efficient network expansion rejected by the AEMC. They suggested a stronger public-sector planning mechanism, with the network companies being much less influential in suggesting network investment priorities. This has been taken up belatedly after the 2016–17 failures. And the Review proposed that where the planner recommended major new investment, its provision and ownership should be put out to tender, so that consumers could benefit from the more cost-effective delivery of systems and the lower cost of capital of alternative suppliers. This was consistent with continued management by established network monopolies.

Transmission has become a major constraint on new investment in renewable energy. Most of the spare capacity in transmission connections from high-quality renewable energy resources to demand centres has now been exhausted. Investment in renewables in the best renewables regions has exceeded transmission capacity, leading to a proportion of each generator's output being stranded from the market ('curtailed', in the language of the industry). Regions with good renewable energy resources had once been importers from the coal regions, and local generators were granted premia (high 'Marginal Loss Factors [MLFs]') for generating power in them. These regions are now exporting power and pay penalties (low MLFs). The changes from premia to penalties have been unexpectedly large and the processes of determining the changes are not well understood. Curtailment and the large adverse changes in MLFs, and uncertainty about how each will change in future, have reduced expected returns from renewables investment. These effects have been compounded by high and unpredictable costs imposed by the transmission and distribution monopolies for connecting new generators. These considerations will seriously constrain new investment in renewables in the period ahead, pending transformational investment in transmission capacity.

Belatedly, expensively and on a small scale, ElectraNet has proposed a new 800 megawatt interconnector between Robertstown in South Australia and Wagga Wagga in New South Wales at a cost of over \$1.5 billion. This has strong support from the South Australian and New South Wales governments. AEMO, in its role as planning adviser on networks to the Victorian government, has proposed expansion of network capacity to connect Melbourne to the high-quality wind and solar resources of western Victoria.

The current transmission and distribution system will not be able to support the vast expansion of demand for power required for the replacement of oil, gas and coal in road transport, industry and household heating, and the utilisation of opportunities to expand output of energy-intensive industry.

To meet Australia's requirements on decarbonisation and opportunities in the future low-carbon world, there will need to be major investment in long-distance high-voltage transmission. We can make a start on our low-carbon opportunity in industry by using the established transmission systems linking the old power-generating regions to many centres of demand, to bring renewable energy back in rather than to take coal-based electricity out. However, the establishment of Australia generally as the natural locus for new energy-intensive industry requires immense quantities of power from outside the established regulated network systems.

I would add a fourth proposal for transmission network investment reform to those that I suggested in 2008 and 2011: reward private unregulated investment that adds public value to the network.

The proponent of new long-distance high-voltage transmission would bear the full cost and the risk, and pay reasonable costs of using the regulated network to take power from the unregulated line to final users. The new policy requirement is to reward private investors for the benefits they provide to the regulated system. Benefits would include increasing reliability and security. AEMO and the AER would rigorously assess the value of the extension of the system to the regulated grid. The rules would be transparent, and applied without the exercise of discretion. The proportion of the cost of new private, unregulated transmission capacity that contributes good value to the regulated system, and only that augmentation of value, would be rewarded as if it were an increase in the Regulated Asset Base in the ownership of the unregulated investor.

The security and reliability crises of 2016–17 prompted much greater interest in pumped hydro storage (PHS). The ARENA-funded study of potential off-river sites led by Professor Andrew Blakers of the Australian National University drew attention to large Australian opportunity for PHS.[5] The Commonwealth government's purchase of New South Wales and Victorian equity in Snowy Hydro has led to announcement of support for 'Snowy 2.0', an additional 2 gigawatts of PHS power.

The Snowy system has 175 hours of storage. That makes it possible to greatly expand the power capacity ten-fold or more if there were value in doing so, while providing abundant hours of storage. An expanded Snowy PHS could meet most of the balancing of solar and wind with close to 100 per cent renewables in the National Electricity Market. It would, however, be expensive. The 27-kilometre horizontal distance between the two storage dams is much greater than is usual in PHS. This raises capital costs and loss of power through friction in operations. Similar potential for balancing mainland renewables at lower cost has been identified in Tasmania. Both the Tasmanian 'Battery of the Nation' and the Snowy 2.0 project would require large investments in transmission to contribute substantial value – the latter through new submarine cables. Other, smaller PHS sites could provide storage at substantially lower total costs per unit of electricity.

ARENA has funded several feasibility studies in South Australia and some in other states for PHS projects in the range of 200 to 400 megawatts. Several PHS projects featured in the Commonwealth's announcement in April 2019 of priorities for Commonwealth underwriting. The Queensland government's new clean energy generator, CleanCo, intends to utilise more fully the large hydro-electric potential at Wivenhoe. Most recently, AGL has announced an intention to proceed with a 250 megawatt PHS project at Hillgrove in South Australia.

Like renewable energy, PHS and batteries comprise mainly capital costs. They have therefore been large beneficiaries of the decline in the cost of capital, discussed in Chapter 3. Chart 4.10 shows that both utility and electric vehicle scale batteries have been experiencing rapidly declining equipment costs with increased installations over recent years. There is no sign of deceleration of cost reductions. Hydro-electricity is a mature technology, so it does not have the same potential for falling equipment costs. If battery costs continue to fall, this would reduce to some extent the economic case for PHS.

But while there is no general downward trend in the cost of mechanical and civil engineering for PHS, there is wide variation in costs between sites. The best Australian sites have lower cost profiles than the averages presented in Chart 4.10.

With gas priced out of long-term balancing of intermittent renewables by the move to export parity pricing in eastern Australia, it is becoming clear that storage will play the major role in providing reliability in the National Electricity Market (NEM). Reliability involves ensuring there is enough supply to meet demand at prices that are not so high as to catastrophically damage the economy through the whole range of variations in supply and demand for power.

Storage will need to be supported by long-distance transmission and demand management. Expanded long-distance high-voltage transmission will reduce the cost of firming renewables by balancing the availability of solar and wind power from different climatic regions.

Chart 4.10 Costs of grid-scale battery and pumped hydro storage (A$2019)

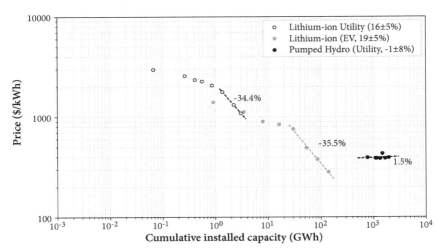

The balance between PHS and batteries will vary with circumstances, with batteries having advantages for fast response applications, including frequency control and for balancing disruptions of short duration. PHS at the best sites will have advantages in shifting multiple hours of generation to availability at another time.

More thought needs to be given to the market and institutional arrangements within which storage and other reliability services operate. The 2017 Finkel Review noted that the energy-only Australian NEM is unusual, and that markets in reserve capacity are common in other systems.[6] AEMO has reiterated those points in various contributions to the public discussion in mid-2019.

The energy-only electricity market that operates in Australia provides incentives for investment in storage, peaking power generation and demand management, through arbitrage between prices at different times. However, the owners of PHS and battery storage or other contributors to grid stability cannot capture the full benefits that they provide to the power system unless they are rewarded separately for that contribution.

The current business model proposed for the Snowy Hydro PHS involves selling hedges commercially into the electricity market. It relies on arbitrage in the energy market. In its 2019 Electricity Market Report, the ACCC expressed concern that if the Snowy's PHS were to proceed, the Commonwealth government–owned Snowy Hydro would own 60 per cent of peaking capacity in New South Wales and Victoria and control pricing in the hedge markets. This would make it impossible for other companies to invest in peaking capacity or to trade with confidence.

A different model is required. I have proposed that the Commonwealth government divides Snowy Hydro into two statutory corporations, separating the established PHS capacity and the feasibility studies of Snowy Hydro from the other operations of the Snowy Hydro generator-retailer. The storage corporation could be called the Snowy

Energy Guarantee (SNEG). The SNEG would be given the sole purpose of guaranteeing the reliability of the NEM. It would be informed by AEMO of the supply and demand outlook for many years into the future, and of risks to that outlook. The COAG Energy Council would specify the upper price limit at which the balancing of supply and demand was to be underwritten. The SNEG would carry the responsibility of ensuring that capacity is always available to balance supply and demand for power in each region at the specified maximum price. It would intervene in the market to secure that maximum.

Within these objectives, the SNEG would be required to minimise the cost of securing reliability. It could choose to utilise the feasibility study for Snowy 2.0, or purchase other storage or peaking assets or services if these would serve at lower costs. It could purchase services from Tasmania's Battery of the Nation, or from other PHS or utility-scale battery or projects. It could secure access to demand management. Purchases of reliability services from thermal generators would not be excluded where these were the lowest-cost means to the reliability end. Through these processes, the SNEG would acquire the capacity to underwrite the reliability of the electricity system as we move to zero net emissions, and into a new era in which new Australian industry greatly expands demand for power.

Electricity as the foundation of the energy superpower

After a dozen years of close acquaintance with the Australian and global energy transitions, I now have no doubt that intermittent renewables could meet 100 per cent of Australia's electricity requirements by the 2030s, with high degrees of security and reliability, and at wholesale prices much lower than experienced in Australia over the past half dozen years.

More importantly, I now have no doubt that with well-designed policy support, firm power in globally transformative quantities could be supplied to one or more industrial locations whenever it is required in each state at globally competitive prices. (That is around $50 per megawatt hour today.) No other developed country has a comparable opportunity for large-scale firm zero-emissions power, supplied at low cost beyond domestic consumption requirements.

If we seize this opportunity, Australia will be the locus of a historic expansion of internationally oriented energy-intensive industry. That will require innovation in policy, business management and application of new technologies that have been developed abroad and in Australia. The innovation will allow us to utilise our high-quality renewables resources, join load and generation centres all over Australia with efficient interconnection, and bring to account a wide range of other opportunities to balance supply and demand for power.

It will take many years to restore globally competitive practices to the established power systems. We can see the slow progress so far in implementing the ACCC's far-reaching recommendations on restoring affordability and Australia's competitive advantage. It will be costly to economic opportunity and our large interests in successful containment of climate change if we wait until our slow-moving business institutions and policy move to correct the errors of the past.

We do not have to wait.

We can let the fleet of foot in business and state and regional government move on their own until the established structures have acted on the opportunity. The three large policy reforms proposed in this chapter will increase the chances of a strong outcome: the extension of the current minister for energy's power investment underwriting proposal to embrace the whole of the ACCC's Recommendation 4 on underwriting firm power sales from new facilities; the reward to unregulated transmission investment for benefits provided to the regulated system; and the securing of reliability through the SNEG. Each could be implemented quickly. Each is fully consistent with the Coalition government's 2019 electoral commitments. Administered well, each would allow innovative businesses to build a globally competitive electricity supply system for expanding Australian industry. Administered well, together they would deliver lower prices, greater security and reliability, and lower emissions in the established system.

Australia emerged as a major global energy player in the later decades of the twentieth century. As we have seen, Australia lost its old domestic advantages in the fossil-energy world economy in the twenty-first century through the internationalisation of local gas and coal markets, and through the privatisation and corporatisation of electricity systems without effective regulation in the public interest.

The legacy of high infrastructure costs will remain with us for some time. We cannot wait for comprehensive reform of the old before building the new. Chapters 5 and 6 point out the potential for expanding Australian electricity demand by more than 200 per cent over the next decade or so to meet the needs of minerals smelting and electrification of transport. Clever facilitation of the new will provide opportunities for reducing the costs from the old – sooner through integration of electric transport, and later from the support that can be provided by new transmission systems built for new industry.

We have a new opportunity for prosperity through building energy-intensive industries for export to the world. Unlike the last time around, success from the new opportunities is sustainable. Chapter 5 expands upon this new opportunity.

Mutually supportive electricity, industrial and transport transformations

Chapters 5 and 6 show that total Australian electricity demand could increase by more than 200 per cent over the next decade or so. Within a sound policy and regulatory framework, this would reduce costs while improving security and reliability of power to established Australian businesses and households. Within a sound framework, interaction with the established, regulated power system would support the industrial and transport transformations.

Within a poor framework, the electricity system would lose these benefits, and the industrial and transport transformations would be delayed, and their full potential never realised. A poor framework would seek large gains for the old system too soon, and block the industrial and transport transformations.

Electricity supply to support large-scale expansion of Australian exports of manufactured products to global markets will rely mainly on new, private, unregulated transmission. This will link high-quality renewable resources to new industrial loads.

The points of interaction between new private and old regulated systems can reduce prices and improve security and reliability for current users of power.

Intelligent pricing of established transmission for servicing industrial demand at old coal generation and minerals processing centres can increase network utilisation and make some contribution to costs.

Intelligent pricing of network connections and use can allow variations in industrial use of power to make large and timely contributions to security and reliability. Turning electrolysers up and down will be of immense value once renewable hydrogen production is established.

Availability of low-cost renewables and storage from the industrial systems will contribute to lower wholesale prices at peak times.

Finally, new sources of supply from geographically diverse sources will reduce the price-setting power of oligopolistic generators in each region.

Chapter 6 shows that electrification of transport could improve or worsen costs, security and reliability, depending on the quality of policy and regulation. The prize for getting this right is very large.

THE INDUSTRIAL TRANSFORMATION

Industry and fugitive emissions now account for about a third of Australia's total. Industry emissions have actually stabilised: confident expectations and then the experience of carbon pricing, together with the contractionary impact of the high real exchange rate during the resources boom, led to the stabilisation of industry emissions over the past decade. In contrast, fugitive emissions (emissions released from coal, gas or oil mining and processing separately from the use of the energy) – in the absence of policy constraints since 2014 – have grown explosively with the expansion of coalmining and liquefied natural gas. The increase in fugitive emissions is the main reason why Australia has failed to make any overall progress towards emissions reduction over recent years, despite falls from electricity generation and changes in land use.

There is an opportunity to change the structure of Australian industry in ways that reduce emissions while also lowering costs and increasing global competitiveness. That would open a new era of expansion in investment and production. Globally competitive electricity prices based on Australia's exceptional endowment of renewable energy resources would provide the opportunity for productive structural change.

Fugitive emissions are harder. Fugitives have contributed virtually all of Australia's increase in emissions since 2014. They will disappear as global use of coal and gas fades away during the twenty-first century. But what happens between now and the conclusion of fossil energy production matters a great deal to Australia's contribution to the global mitigation effort over the next several decades. The timely reduction of fugitive emissions will require regulatory intervention in the absence of a carbon price. The Commonwealth, through the safeguards mechanism of the Emissions Reduction Fund, is in a position to ease the problem; there are also instruments for intervention in the hands of state governments.

With regulatory support, fugitives would be substantially reduced at relatively low cost through geosequestration and relying less on resources with high emissions profiles. The balance of the reduction could come from purchase of Australian Carbon Credit Units (ACCUs) generated in and sold from the Australian land sector.

Recent developments

In 2008, I concluded that an effective global mitigation effort would adversely affect the competitiveness of Australian energy-intensive industry in the short term. This is because mitigation would tip the balance in favour of zero-emissions energy and away from fossil energy as an input to industry. At that time, we had a strong comparative advantage in fossil energy, but not in renewable energy. For a while, new global investment in energy-intensive industry would mainly occur near favourable low-cost renewable energy resources – including under-utilised hydro-electric and geothermal potential in developing countries. The 2008 Review expressed high confidence that Australia would again become an attractive location for energy-intensive manufacturing (for example, aluminium smelting) in a low-carbon world, but not for a couple of decades, after the exhaustion of underutilised hydro-electric capacity.

Chart 5.1 Industry emissions 1990–2019

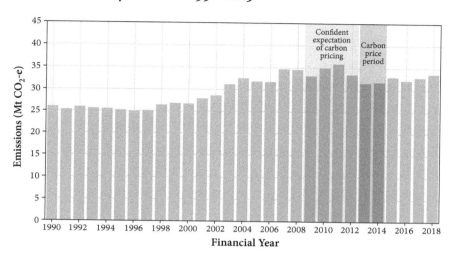

Chart 5.2 Fugitive emissions in industry 1990–2019

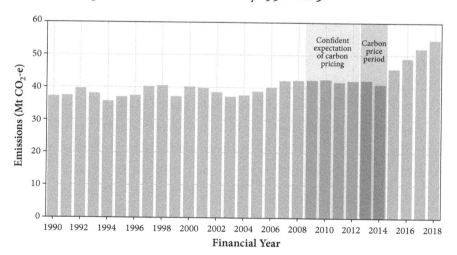

For industry, the fall in the cost of renewable electricity from wind and solar has brought forward by over a decade the time when Australia's rich renewable energy resources provide this country with decisive competitive advantages in energy-intensive manufacturing. There has been major progress over recent years in Australia and globally in thinking about how we can decarbonise industry at low cost. This has proceeded productively at our main public institutions for research in the technological and engineering sciences, including the CSIRO and the school of engineering at our major universities. ClimateWorks Australia and Beyond Zero Emissions have played important roles in synthesising relevant knowledge and raising public understanding of what is possible. More recently, the Australian German College and the Climate and Energy Transition Hub, based in Australia at the University of Melbourne and the Australian National University, have played important roles. These institutions have plugged Australia into a massive global effort to understand and promote the global transition to a low-carbon economy. There has also been considerable progress globally but not yet much in Australia in the development and commercialisation of low-carbon manufacturing technologies.

Australian manufacturing industry enjoyed extraordinary dynamism during the reform era from 1983 to the beginning of the twenty-first century. The volume of manufacturing exports increased at double-digit rates for two decades. It stagnated in the decade or so from the early 2000s. Over the next several decades, there is an opportunity for unprecedented dynamism and expansion in production and export of manufactures that use electricity and biomass intensively. The scale of the expansion would go well beyond compensating for declining fossil energy exports.

Decarbonisation of global manufacturing will build on electrification supported by the replacement of fossil by renewable electricity, hydrogen produced by electrolysis from renewable electricity, biomass replacing fossil hydrocarbons as industrial raw materials, and the capture and storage of emissions in processes in which fossil carbon inputs are more expensive to remove. There will be competition between zero-emissions technologies for processing Australian ores – for example, for reduction of iron ore to iron metal, between hydrogen-based and coke-based reduction of iron ore with carbon capture and storage (CCS). The outcome will depend on the speed at which costs are reduced with increased deployment of the new technologies.

Competitiveness in the zero-emissions economy will be assisted by improvements in energy efficiency and recycling. Australia has been mostly weak with energy efficiency in industry. We need to catch up with other developed countries and China. We have a mixed record on recycling, and need to be unambiguously strong. Improving our performance in energy efficiency and recycling is important for reducing our emissions as well as for our economic performance.

In Chapter 4, I showed that Australia could have globally competitive electricity in a zero-carbon world economy. Low-cost electricity is the main determinant of competitiveness in hydrogen-based processes. Australia has much more than its per capita share of potential for industrial use of low-cost biomass. It also has more than its share of low-cost CCS opportunities.

The Energy Transition Commission noted that the lowest-cost renewable power available for large-scale industrial use today is about US$35 (about A$50 at mid-2019 exchange rates), in locations with exceptionally favourable resources, in Iceland and parts of Canada.[1] Australian renewables-based generation from good resources could be firmed and delivered to favourably located industrial centres at these costs within the policy frameworks suggested in Chapter 4. This could be done now, with the electricity infrastructure being built within the time it takes for construction of new industrial capacity.

Australia's competitiveness in a zero-carbon world

Australia is more richly endowed than any other developed country relative to population and local demand for both fossil and renewable energy. However, Australia's international competitiveness in electricity supply will strengthen as all countries move away from fossil energy in electricity supply. This is because fossil energy can be imported at relatively low cost by countries with poor resources, but renewable energy cannot.

Thermal coal does not cost much more in China or Japan than in Australia. However, renewable energy can be traded internationally only at high cost – through high-voltage direct-current submarine cables, or through liquefaction of hydrogen or use of ammonia as a hydrogen carrier. International transmission of electricity from Australia would cost much more than the electricity itself. More than half of the value of hydrogen would be lost in preparation for and during transportation. Unlike Australian coal resources, Australian renewable electricity will be available at much lower cost to Australian than overseas industry.

Australian power generation and industry once had access to Australian fossil energy resources at prices far below those in importing countries. Until early this century, high-quality Australian coal resources were reserved for use by the electricity commissions of each of the eastern states, and coal made available at cost to them. However, coal in Queensland and New South Wales is now exportable, and power generators are required to pay export parity prices for them – not much below prices in countries importing Australian coal. Until a few years ago, eastern Australia had domestic gas prices at a modest fraction – at times around a third – of those in North America, Europe and Japan. The development of the liquid natural gas (LNG) export facilities in Gladstone lifted Australian gas prices to export parity levels – not much below and at times equal to or above those in Japan and Europe. In the meantime, the United States – having much higher domestic gas prices than Australia until a few years ago, and experiencing a proportionately smaller expansion of domestic gas reserves and production to that which in Australia fuelled the LNG export boom – restricted gas exports and domestic prices fell to about a third of the new, higher Australian levels.

An advantage lost and gained

Transport costs of mineral ores add to advantages from low renewable electricity costs to encourage the processing of Australian mineral resources at home. These advantages are present whether the processing uses fossil or renewable energy. Together with the low cost of energy, they made Australia the world's main exporting country for aluminium metal in the 1980s and early 1990s. As Australian wholesale energy costs rose to those in other developed countries – and beyond – and with extraordinary increases in domestic energy distribution costs, we lost our international competitiveness in aluminium smelting and other energy-intensive industry.

The overall advantages of processing Australian mineral ores in this country are much greater in the zero-emissions world economy than they were in the coal and gas era. While the energy resource advantages for Australia may be comparable in the fossil and renewable energy economies, there are much larger transport cost advantages in using electricity at home in the new era.

Coal is transported from Australian to northeast Asian ports for around 10 per cent of its value. Natural gas turns from gas to liquid at –160°C and absorbs about one-tenth of the resource in liquefaction. Liquefaction requires heavy capital expenditure. Its shipping costs almost as much as liquefaction. Regasification adds another layer of costs that is only partly covered by heat recovery in this process. Hydrogen liquefies at –253°C – close to absolute zero. Liquefaction absorbs about 30 per cent of the calorific value of hydrogen and requires much greater capital expenditure than natural gas. Because hydrogen liquid has to be kept at much lower temperatures than LNG and is much more voluminous than natural gas, and because its small molecules are more demanding on materials for containment, shipping is substantially more elaborate and expensive than for natural gas. Total costs of hydrogen at ports of destination are likely to be more than twice as high as in locations of origin in Australia. Long-distance high-voltage direct-current submarine transmission of electricity to Java and through Indonesia to the mainland of Asia would also more than double costs at the place of delivery from Australia.

It follows that the economics strongly favour the use of Australian electricity and hydrogen to process Australian minerals at home, rather than sending both the raw materials and the hydrogen or electricity to Asian destinations for conversion into metals. This advantage from low-cost domestic energy is much greater than in the gas and coal era, when energy could be made available to industrial plants overseas at costs much closer to prices at home.

Separately, and beyond savings in transport of raw materials mined in Australia and lower energy costs, Australia has several advantages in processing many minerals. These processes are capital-intensive, and outside the many industries in which oligopoly artificially elevates costs, Australia, like other developed countries, generally has access to capital near globally competitive prices.

Australia also has rich resources of human capital across the range of expertise required for globally competitive large-scale processing of resources: engineering, financial, geological, metallurgical and project management.

The low-hanging fruit will be the processing of Australian materials that uses electricity intensively (either directly or as hydrogen), or carbon from biomass, or that requires the capture of carbon emissions. The advantages are magnified when more than one of these advantages are relevant.

We also have disadvantages in high construction costs (especially in northern Australia) and in many transport, energy distribution and other infrastructure services. Our big disadvantage is oligopoly, high costs and inflexibility in supply chains, including infrastructure services. These have been analysed in recent reports of the Productivity Commission and the ACCC. It is not easy to break apart deeply entrenched anti-competitive structures and habits. The early stages of realisation of the Australian opportunity may require new entrants with more positive attitudes to innovation and globally competitive pricing.

Australian energy-intensive processing industries can be globally competitive when the transport cost savings on raw materials, lower energy costs and Australian skill advantages outweigh the drags on our competitiveness. They did outweigh them in aluminium and many other energy-intensive manufacturing industries for a period late last century. They ceased to do so until now in the twenty-first century as energy prices rose to and in some case above international levels, and with the high real exchange rate during and immediately following the resources boom. There is an opportunity to regain competitiveness and for it to increase far beyond any early experience in the emerging low-carbon world economy.

There are many large industries in which the price of electricity is or could be a major factor in global competitiveness.

Manufacturing aluminium metal competitively

Aluminium is currently the most electricity-intensive product entering world trade in large volumes. At market prices for aluminium metal and $50 per megawatt hour of electricity, electricity accounts for about one-fifth of the cost of the product. China produces more than half of the global annual total of about 60 million tonnes, nearly all for its own use. China's production grew rapidly through the first twelve years of the twenty-first century, pushing down global prices. In recent years, within the new model of Chinese economic growth, China has downgraded the priority of investment in new aluminium capacity, as its energy intensity and environmental impact do not fit China's post-2012 model. As the growth in world demand for aluminium requires new capacity somewhere in the world, it is likely to be placed in the most competitive locations outside China.

Australia is a modest global producer of aluminium (about 2.5 per cent of the total) but a substantial exporter (nearly 10 per cent). Australian production grew on low-cost hydro-electric power in Tasmania. It expanded on a much larger scale with low-cost coal in Queensland, New South Wales and Victoria as world energy costs increased and the priority of environmental amenity rose in Japan in the late 1970s and 1980s. Japan for a while in the 1970s was the Western world's largest producer of aluminium. It shifted swiftly to being the biggest importer in the 1980s, with much of its requirements being drawn from new smelters in Australia. Australian competitiveness in coal-based aluminium smelting has been challenged by rising electricity costs in the early twenty-first century, and no mainland Australian smelter is at present certain of long-term survival in the absence of fundamental changes in electricity supply.

Aluminium is smelted from alumina – refined aluminium oxide from the natural mineral bauxite. Australia is the world's largest producer of alumina (about a quarter of the world's total), most of it exported to smelters overseas. Australian production occurs at two locations. The bigger is between Perth and the southwest towns of Collie and Bunbury. The other is Queensland, with bauxite from the Gulf of Carpentaria being refined into alumina at Gladstone in Central Queensland.

Aluminium smelting in Australia is responsible for large amounts of carbon dioxide emissions. The main source of Australian aluminium emissions is the generation of power from coal. There are also large emissions from within the process itself, from oxidation of the fossil-fuel-based carbon cathodes and anodes used in electrolysis. The two global aluminium companies smelting metal in Australia, Alcoa and Rio Tinto, have committed to producing the metal globally with zero emissions. This could involve supply of cathodes and anodes for electrolysis made from non-carbon materials, or from biomass.

There are substantial cost advantages in smelting aluminium adjacent to alumina refining. This is only economically feasible if globally competitive power is available at those locations. This is achievable now. In the absence of artificially high costs for other inputs into production, Australia is naturally the home of new large-scale smelters. The Australian advantage will strengthen as the world moves towards zero-emissions electricity, causing costs to rise in other countries but not in Australia. Smelting half of Australia's alumina exports at home would require construction of four or five world-scale plants, increase aluminium production severalfold, and increase Australian electricity demand by around a quarter Aluminium-smelting technologies currently applied in Australia are interruptible for about one hour in twenty, so that their presence is useful in stabilising an electricity system against short-term shocks. Plants can now be designed with greater flexibility, expanding their value to system reliability.

Manufacturing clean iron competitively
The largest opportunity is in production of iron metal or steel from iron oxide ores. Australia is by far the world's largest producer (nearly two-fifths of the world total) and exporter (nearly three-fifths) of iron ore. China takes about 70 per cent of Australian exports. China has supplied most of the increase in global steel production in the twenty-first century, and now accounts for around half of the global total output of about 1.8 billion tonnes.

Much of the steel produced in the old developed countries in North America, Japan and Western Europe is now from recycled scrap, through the electricity-intensive electric arc process. This can be zero-emissions steel if it is powered by renewable electricity. But China and other developing countries do not yet have the steel consumption legacy that produces large amounts of scrap, and most of their output is from highly emissions-intensive reduction of iron ore using coke. The waste from removing oxygen from iron ore to iron metal is carbon dioxide. The production of iron using this process is by far the world's largest industrial source of carbon emissions. It contributes about 7 per cent of global emissions.

The lowest-cost lower-carbon route currently used commercially for converting iron ore into iron and steel is direct reduction using natural gas. The direct reduction process is now well established. The most widely used process, Midrex, is jointly owned by Japanese (Mitsubishi) and German (Siemens) engineering companies. Other processes owned by European companies have advantages in particular circumstances, including where high proportions of hydrogen are injected into the gas mix. Direct reduction using natural gas reduces the ratio of carbon dioxide to iron produced by about half of that in the blast furnace. About 75 million tonnes per annum of iron metal are now produced globally through direct reduction.

The direct reduction process uses the reduction properties of both hydrogen and carbon in the gas. The process works as well if extra hydrogen is added. Increasing proportions of hydrogen are being introduced with natural gas in some plants. Up to 70 per cent hydrogen has been substituted for natural gas without changes in the plant. This reduces emissions to 15 per cent of blast-furnace production. Some plants have operated with 100 per cent hydrogen and zero emissions.

The iron metal from direct reduction using natural gas or hydrogen is converted into steel in an electric arc furnace. If hydrogen made by electrolysis using renewable electricity is used to reduce the ore, the electric arc that converts iron metal into steel is powered by renewable energy, and the relatively small carbon inputs necessary for the process are drawn from biomass, the primary steel is produced with zero emissions. This now looks to be the likely route to removing emissions from steel-making. The main alternative is reduction with coke in the blast furnace, accompanied by capture and storage of carbon dioxide emissions. CCS has high costs, which can be expected to fall with increased deployment. This may be competitive with the hydrogen route in highly favourable locations adjacent to high quality geosequestration sites.

The hydrogen and electric arc route to steel is one of the most electricity-intensive major industrial processes. Each tonne of zero-emissions steel requires nearly 5 megawatt hours of electricity. The electricity component at A$50 per megawatt hour represents about 25 per cent of the value of crude steel products (say, steel slabs or hot rolled steel coil, for export and further processing in importing countries).

Western Australia has a different gas policy than eastern Australia. Since the commencement of LNG exports, it has reserved a proportion (15 per cent) of output for domestic use. This has insulated Western Australia from the huge increases in domestic gas prices in eastern Australia that have accompanied the emergence of exports from Queensland. Gas prices in Western Australia are now about a third of those in the east. This makes Western Australia the natural Australian home for industrial processes requiring use of natural gas, pending the emergence of zero-emissions alternatives.

Where new plants are required, direct reduction with natural gas and conversion into steel in an electric arc furnace is highly competitive with blast-furnace reduction using coke at mid-2019 WA gas prices if power is available at $50 per megawatt hour. Hydrogen is not competitive with natural gas in Western Australia at today's domestic gas prices and today's cost of electrolysis plants to make hydrogen from water and renewable electricity – even with electricity at $50 per megawatt hour. Commercial use of a proportion of hydrogen in the direct reduction process in Western Australia now would require recognition of the value of reduced emissions or direct capital support for the electrolysis plant to produce hydrogen.

In the zero-emissions economy, the relevant comparison is between direct reduction of iron ore using hydrogen, and the coal-based blast-furnace route with CCS, The reduction of iron oxide (high-grade iron ore) requires about 51 kilograms of hydrogen or 450 kilograms of carbon in the form of coke. At mid-2019 world prices, coke costs about A$150 per tonne of steel. That means that hydrogen would need to be available at less than A$3 per kilogram for the cost of the reductant to be similar in the two processes. BloombergNEF reported in August 2019 that current costs globally were in the range A$3.75–A$10, depending on scale, source of electrolysers (much cheaper from China) and cost of electricity. It expected costs to fall to A$2.90 by 2030. That presumed combinations of wind and solar power without firming (i.e. the electrolysis plant ceasing production when neither the solar nor wind energy was available) at A$36 per megawatt hour in 2030. Costs for non-firm solar and wind close to these are already available in the best Australian locations.

Many costs other than those of hydrogen and coke will determine the outcome of competition between hydrogen-based direct reduction and coal-based blast furnaces with CCS. At present, the high cost of CCS would place the blast furnace a long way behind.

Costs of CCS vary widely with proximity to favourable geological structures: the best may be competitive even if these provide opportunities for production of only a small proportion of the world's primary steel. Electrolysis plants for producing hydrogen and CCS costs will both fall a great deal with larger scale of deployment. Hydrogen is more likely to get that boost from scale earlier, as demand will increase from transport, electricity generation and other uses besides reduction of iron ore. The capital requirements are currently similar for the two processes. The hydrogen route produces iron with fewer impurities, which makes it more suitable in some high-value uses, and gives it advantages for mixing with scrap in the electric arc. Electrolysis for making hydrogen produces eight times the mass of oxygen as of hydrogen, so that location in an industrial region with demand for oxygen reduces the cost of producing hydrogen.

With prospects of large cost reductions in electrolysis plants for hydrogen, and further cost reduction for renewable energy, the costs of using West Australian natural gas and hydrogen may be comparable in the 2030s – depending on what happens to natural gas prices. With recognition of the value of zero emissions, cost parity would come much earlier.

Meanwhile, increasing carbon constraints will raise the cost of the blast-furnace route to iron-making. Investors who are aware of carbon risk would not invest now in new blast-furnace iron-making without having access to globally competitive CCS opportunities. There is now a strong commercial case for building new primary iron-making capacity as direct reduction facilities using natural gas near iron ore mining in Western Australia, initially mainly using natural gas, and increasing the renewable hydrogen proportion over time. Initially, the hydrogen component would require public support for innovation, pending the expansion of the scale of electrolysis and the associated reductions in cost. Alternatively, carbon pricing in the markets for the iron and steel (most importantly in Europe and northeast Asia), or regulatory restrictions on use of high-emissions steel, or development of 'green' premia for low or zero-emissions steel would bring forward the time at which hydrogen-based iron-making is fully competitive with conventional processes.

As the world moves closer to requiring zero-emissions steel, there is scope for iron metal and steel production at an immense scale near

Australian iron ore resources in the Pilbara, the mid-west and south-west of Western Australia and in the Upper Spencer Gulf and adjacent parts of South Australia. We can expect that in a zero-emissions world, iron production will migrate to Australia from those parts of industrial Asia with the weakest endowments of renewable energy resources. Japan and Korea would be first. A proportion of Chinese capacity would follow. Only a proportion from China, because that country has greater renewable energy resources than its northeast Asian neighbours, although a small fraction of Australia's in comparison with domestic demand. These northeast Asian countries are at or past peak steel consumption and production, and the proportion of domestic requirements supplied from recycling scrap is high (Korea and Japan) or will grow rapidly (China). Migration from Europe would occur with free trade in and iron and steel and acceptance of imported green steel in border price adjustments related to carbon. For developing countries still experiencing rapid growth in primary steel demand, most importantly in southeast Asia (first Indonesia) and south Asia (first India), the low-cost route to zero-emissions supply of steel would be to import high proportions of iron metal or steel from Australia for rolling and final production of manufactured goods near the domestic market.

Reducing a tenth of the iron ore exported from Australia into crude steel through renewables-based hydrogen would roughly double Australian total electricity demand. Both electrolysis for making hydrogen and the electric arc for turning iron metal into steel are interruptible processes, so that a large presence helps reliability in the whole electricity supply system.

Other energy-intensive manufacturing
Production of pure silicon from sand or quartz and its incorporation into other valuable compounds is one of the most energy-intensive industrial processes. In the one Australian plant, owned and operated

by Simcoa Operations at Kemerton in the Collie–Bunbury region of Western Australia, about one-third of costs are energy and 30 per cent carbon. Carbon is drawn in the ratio of 2:1 from biomass and fossil sources. Demand for high-grade silicon has increased rapidly with production of computers and PV panels – in both of which pure silicon is the critical material input. Uses of output from particular plants are constrained by purity, with high-grade product depending on the quality of the sand or quartz and of the carbon (fossil or biomass) used in the process.

Australia is a small silicon producer in the global context, but commands interest and high prices for exceptional quality. China produces about two-thirds of global silicon. Chinese production is challenged by the high costs of electricity (much higher than the globally competitive prices potentially available in Australia) and by local environmental constraints. Chinese production has not increased for several years, and China's large export volumes have been absorbed in increasing proportions by growing domestic demand. The economically efficient development of the global market would see expansion of production in places with access to globally competitive power and adjacent to supplies of high-quality quartz and sand. Several Australian locations meet these criteria.

Ammonia plays a large part in the global production of nitrogenous fertilisers, explosives and some other chemical products. Global production of around 150 million tonnes per annum is overwhelmingly from fossil fuels. Hydrogen is extracted from gas or coal and combined with atmospheric nitrogen under great pressure and heat through the Haber–Bosch process. The production of hydrogen in this way is highly emissions-intensive and the Haber–Bosch process highly energy-intensive. The latter can have zero emissions through the use of renewable electricity as the energy source. The production of hydrogen could be with zero emissions if hydrogen were produced through electrolysis using renewable electricity. Renewable ammonia production is one of the most energy-intensive industrial processes and would gravitate economically towards low-cost renewable energy in Australia.

There is strong demand for renewable ammonia in Europe in particular, and increasingly in Japan and Korea. Beyond its current uses, it is potentially a carrier for hydrogen, which in some circumstances can be shipped at lower total cost than hydrogen in liquid form. A premium can be earned for 'green ammonia', which is helpful to launching hydrogen-based projects.

Australia currently contributes only about 1 per cent to global ammonia output, mostly for domestic fertiliser and explosives. There are exports from the Yara plant in the Pilbara, which is owned by the Norwegian sovereign wealth fund, whose owner has said that it must shift to renewable hydrogen within a reasonable period. A hydrogen pilot plant is being developed to supply the ammonia plant with part of its reqirements. Australia is at the forefront of developing zero-emissions ammonia from renewable electricity, with another pilot plant having been announced for Port Lincoln in South Australia.

The conversion of renewable hydrogen into ammonia is through a standard process. The change that is required to launch renewable ammonia on a large scale is reduction in the cost of electrolysis plant and therefore of the cost of hydrogen for the Haber–Bosch process. This will be driven by rapid expansion of global hydrogen demand for transport, steelmaking and other industrial processes, as well as for ammonia. With perhaps a thousandfold increase in global hydrogen production required for realisation of zero global emissions in industry, power generation and transport at the lowest possible cost, there are prospects for large reductions in electrolysis costs, along similar paths to those followed by solar PV and wind generation and battery storage. Together with the anticipated reduction in renewable energy costs, there are prospects for renewable hydrogen becoming competitive with fossil energy hydrogen over the next decade or so. Earlier adoption would be encouraged by grants for innovation, or a carbon price that recognises the zero-emissions production process. Scale in Australian domestic demand, as well as global demand, is important in driving down the domestic costs of hydrogen production.

Biomass as a base for industry

Biomass is the carbon material accumulated in plants and algae as photosynthesis converts solar energy and atmospheric carbon dioxide into hydrocarbon solids and oxygen. It was originally the source of the fossil energy which has powered modern economic development over the past quarter of a millennium. It is a renewable source of energy or industrial materials if the plants from which it is drawn are continually replenished.

Biomass will emerge as a scarce and valuable base for the chemicals industry in the zero-emissions global economy. It will be driven to its most valuable uses by high prices. For some time, biomass will be required for fuels for long-distance civil aviation. Some of the high-value uses will be as inputs into energy-intensive processes. Carbon fibre is one of these. There are many others. There will be special opportunities in Australian regions which have access to low-cost renewable electricity and biomass. These include the southwest of Western Australia, southeast South Australia, the Riverland and Sunraysia regions on the Murray River straddling the South Australia/Victoria border, the Riverina in New South Wales, southwest Victoria and the Latrobe Valley, and the sugar coast of Queensland, centred on Mackay. Where these locations are adjacent to high-quality geosequestration sites, as in southwest Western Australia and Victoria, capture and storage of emissions will lead to the holy grail of climate change mitigation: negative emissions through bioenergy carbon capture and storage (BECCS).

Locations of zero-emissions industries

Initially, the Australian advantage in processing materials will be strongly focused on a few regions. For processing of minerals and metals, it will be an advantage to be located close to mining of the raw material. For aluminium, that will be especially southwest Western Australia and Queensland at Gladstone and further north. For both iron and aluminium, there will be opportunities for transporting oxide ores to particularly favourable processing locations. For iron, that will be Western Australia, with its domestic reservation of gas and adjacent opportunities for CCS. Later, when the world has moved to zero-emissions iron, the Upper Spencer Gulf and adjacent Eyre Peninsula region of South Australia will have advantages. Silicon opportunities will be more widely spread, through the many locations with high-quality quartz or sand. The smelting and refining of other metals, including copper, nickel, titanium, cobalt, vanadium and lithium, among the elements for which demand will be boosted exceptionally by the zero-emissions transition, will be undertaken at strong industrial locations at reasonable distance from the mines. Many energy-intensive chemical processes will have special advantages close to sources of biomass. Some products have only electricity as a major material input, and can be located wherever electricity costs are low and industrial facilities of high quality. Ammonia and the nitrogenous fertilisers and explosives made from it provide the most important example.

New industrial strengths will be built more easily in provincial cities with strong industrial traditions, and established energy, port, other transport, and training infrastructure. This points to the Collie–Bunbury region in Western Australia; the mining ports of the Pilbara; the Upper Spencer Gulf in South Australia; Portland and the Latrobe Valley in Victoria; Port Kembla and Newcastle in New South Wales; Gladstone, Townsville and Mackay in Queensland; and the established materials processing regions of Tasmania. There will be opportunities in some of these locations to break free from the Australian historical legacy of high electricity costs through provision of mainly unregulated electricity transmission from world-class renewable energy resources.

There are advantages in location at the transmission nodes built around the declining fossil power generation: Collie; the Upper Spencer Gulf; the Latrobe Valley; Newcastle; and Gladstone. Northern Tasmania has a similar advantage from the hydro-electric legacy. Here, the established electricity networks can be used to bring low-cost renewable electricity back to the industrial centre, rather than to transmit coal-based power out to other regions. Portland is adjacent to Heywood, which has well-established transmission infrastructure. Northern Tasmania's transmission strengths would be enhanced by greater interconnection with Victoria across Bass Strait. The advantages are greater if there are low-cost hydro or pumped hydro storage sites nearby: the industrial cities close to the Great Dividing Range (the Latrobe Valley, Port Kembla, Newcastle, Gladstone, Mackay, Townsville); Tasmania; and deep depleted mines.

The biggest challenge in industrial use of biomass is usually the high cost of collecting dispersed material. This problem has been solved, although not always at low cost, for the paper mills and wood-chip export facilities. Large amounts of biomass are available as bagasse from sugar processing along the coast of central and north Queensland. Research, development and commercialisation of biomass concentration technologies hold out prospects for low-cost transportation of processed carbon and hydrocarbons to major industrial plants. Australia more than any other country has prospects for low-cost production and harvesting of biomass from land that is not suitable for food production – for example, from the Mallee across the drier parts of the West Australian, South Australian and northwest Victorian wheat-belts, and the mulga country of New South Wales and Queensland.

Algae is potentially an important biomass feedstock to industry in many coastal locations.

The realisation of value from these large opportunities requires research, development and commercialisation effort in Australia and globally. Public support for innovation is necessary, as private investors are unable to capture most of the benefits. This was the rationale for establishment of ARENA, to support cost reduction through support for innovation in renewable energy. Similar mechanisms are now required to support innovation in zero-emissions industry.

The global mapping identifies Australia as a region of exceptional opportunity for low-cost geological sequestration. The best sites have geological structures that are favourable for permanent storage, well-known from oil and gas production and close to major industrial centres with access to low-cost electricity and sources of emissions. Globally competitive sites include the Harvey River Basin adjacent to Collie–Bunbury, East Gippsland and the Bass Strait adjacent to the Latrobe Valley, and the Otway Basin adjacent to Portland.

The Australian low-carbon opportunity can build on old provincial industrial centres with multiple favourable characteristics. Success in these locations will support transformation of energy supply at a later stage in the major cities and through the areas covered by the established regulated energy networks.

Fugitive emissions

Fugitive emissions are of large immediate but transient significance. They will disappear with coal and gas mining by around the middle of the century, except to the extent that coal and gas use in Australia and abroad is broadly reconciled with climate stability by carbon capture, utilisation and storage (CCUS). But they do matter currently, because their recent growth has been large and fast enough to knock off course Australia's delivery on its international commitments.

Fugitive emissions come in several shapes, of which three are particularly important: carbon dioxide co-mingled naturally with methane and hydrogen in natural gas and released in gas processing (the largest and most rapidly growing element); carbon dioxide released from natural gas combustion in gas processing; and methane released from the mining of coal.

Carbon dioxide released in gas processing increased by over 7 per cent per annum from 2005 to 2017, up from a touch over 2 per cent over the previous decade and a half.

There are huge variations in emissions across gas fields. Many fields have up to 25 per cent carbon dioxide by volume (80 per cent by weight). In addition, about 10 per cent of gas energy is typically consumed in processing and liquefaction. In the absence of CCS, well over a tonne of carbon dioxide can be released into the atmosphere for every tonne of gas landed in Asian markets. While the burning of gas in a power generator may only produce around half of the carbon dioxide emissions from burning coal, the replacement of coal by gas from these fields may not always contribute to global emissions reduction.

Methane emissions from mining and gas production and transport also vary greatly across mines. They are poorly measured. Improved measurement is necessary for Australia to meet its obligations. Poor measurement will not permanently prevent detection of the emissions realities, as new remote sensing is becoming increasingly reliable.

There has been recent debate in Australia about whether we should take responsibility and accept credit for reducing downstream emissions from combustion abroad of Australia's fossil-fuel exports. Environmental activists have argued for inclusion of 'Scope 3' emissions – those generated downstream from fossil-fuel mining, whether or not the combustion occurs in another country – in Australia's mitigation responsibilities. This has been most contentious in relation to coalmining, where opposition to development of the Galilee Basin has focused on emissions from use of the coal in other countries. The natural gas export industry has argued for less onerous domestic mitigation contributions because exports reduce emissions through displacement of coal combustion in other countries.

While it is worth discussing these issues, we should be clear on the content of current international agreements on these matters. Each country is responsible for emissions within its own territory, and other countries for emissions within theirs. We do not take responsibility for emissions from burning of Australian coal and gas, or credit for replacing a more by a less emissions-intensive fuel. But there is no avoiding full accountability for Australian fugitive emissions, even if they derive from gas exports that may play a role in reducing emissions abroad.

Other countries' mitigation now assists the gas export industry in Australia by increased demand for gas and higher prices on world markets.

The good news is that most fugitive emissions can be reduced at low cost. The West Australian government had regulations in place a decade ago mandating the capture and storage of fugitive emissions from new LNG projects. This encouraged the reinjection of carbon dioxide into the geological structures from which the gas had been removed. Capture costs are low because the carbon dioxide streams are relatively pure as they emerge from processing plants. Storage costs are low because favourable geological structures are nearby.

These regulatory conditions did not deter large expansion of LNG production in and export from Western Australia. They had no effect on incentives to invest at all in the many large energy companies, including Chevron, operator of the huge Gorgon project, which imposed its own internal shadow carbon pricing in implementation of good practice and anticipation of future mandatory requirements.

The bad news is that low-cost opportunities for stopping the increase in fugitive emissions through CCS in the gas export industry have been missed as a result of incoherence in Australian climate policy.

After the Clean Energy Future federal legislation in 2011–12, the states (sensibly) repealed most of their mitigation policies. The West Australian government's October 2012 climate change strategy, released a few months after the commencement of carbon pricing, announced that the state would, from that time forward, consider climate change mitigation a responsibility of the Commonwealth. Western Australia removed regulatory requirements for CCS on LNG projects.

Following the Abbott government's repeal of carbon pricing in 2013–14, the LNG and coal industries faced no restraints at all on fugitive emissions. The scene was set for the explosion of emissions. No matter how low the costs of mitigation, it is never cheaper to capture and store polluting gases than to release them into the atmosphere.

So what do we do now? There is no doubt that the LNG and gas producers which generate fugitive emissions impose a climate cost on the citizens of the whole world, including Australians. There is no doubt that Australians in general will have to pay for the increased fugitive emissions, through larger reductions of emissions in other sectors or the purchase of offsets. There is a case for placing those costs on the sources of emissions: purchase offsets for any emissions that are not captured or otherwise avoided by management of mining. The costs of avoiding fugitive emissions or of payments for offsets would represent a modest proportion of total revenue and profitability in most coal and gas mines.

As Australians realise the many advantages of participating fully in the global transition to zero emissions, pressures for policy reform will build. This will support an industrial transformation that is being driven by economic imperatives.

To resolve the problem of fugitive emissions, someone must pay. Sound policy will ensure that it is not the ordinary Australian citizen. Paying by purchasing domestic offsets could be important to sustainable transformation of land use. Some contribution has been required through application of the safeguards mechanism of the Abbott government's Emissions Reduction Fund. I come back to that in Chapter 7.

THE TRANSPORT
TRANSFORMATION

Transport emissions have risen steadily for several decades and now represent about a quarter of the Australian total. Transport was not covered by carbon pricing, so, unlike most sectors, there was no break in trend from 2008 to 2014. Yet transport emissions are likely to fall as Australia moves to electrification of road transport and greater use of public transport relative to private cars.

Battery and hydrogen electric vehicles will sooner rather than later be highly competitive with and replace internal combustion engines. The expansion of rail and the development of more compact cities with greater opportunities for public and active (walking and cycling) transport are valuable in themselves, but also important to emissions levels for the period before the full electrification of motor transport.

A follower, not a leader

Australia has established itself as a slow follower in the global transport transition. Other countries are developing and applying the low-carbon transport technologies of the future. While European and high- and middle-income Asian countries have been establishing a dense network of fast electrified rail, Australia has mostly been living off the legacy of nineteenth- and early-twentieth-century heavy and light rail investments. Australia is late in preparation for and investment in electric road transport.

However, since the main zero-carbon transport technologies, from road vehicles to trains and short-haul domestic air travel, all use electricity as a fuel, Australia's renewable energy cost advantages will eventually flow through to transport. Although a laggard now, once the Australian adoption of electric battery and hydrogen road motor vehicles begins, it is likely to move quickly.

The transport transformation is built on the electrification of vehicles and the decarbonisation of electricity; hydrogen-based processes built on renewable electricity; the substitution of biofuels for fossil energy in uses where batteries and hydrogen fuel cells do not yet provide a sufficiently high ratio of energy to weight; and expansion of the rail infrastructure.

The industry and transport transformations are mutually supportive. Expanding hydrogen use in both transport and industry helps to provide scale, which brings costs down for both sectors. Similarly, with a sound policy framework, the interaction of electrification in the industry and transport transformations can reduce costs through more balanced and more complete use of the electricity grid.

The future of transport is overwhelmingly electric. The electrification of transport will either come directly (for railways with sufficiently intense usage to justify the infrastructure) or indirectly. The indirect supply of electricity will come through storage of renewable energy in batteries or hydrogen. Batteries and hydrogen fuel cells will battle for competitiveness across the range of passenger and freight road vehicles. Ammonia as a carrier for hydrogen may turn out to be a low-cost route to zero emissions for long-distance shipping. More efficient aircraft are likely to become zero-emission by using batteries over short hauls and biofuels over long distances.

The balance between battery and hydrogen electric vehicles will

Chart 6.1 Transport emissions 1990–2018

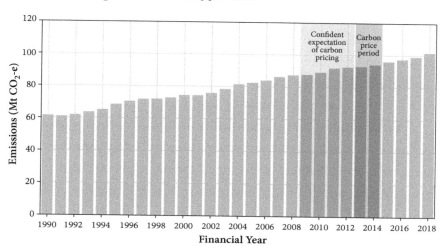

be determined by the relative rates at which costs fall for the electricity engineering systems, by the rate of increase in range of hydrogen fuel cells and batteries (with the former currently having the edge), and also by the rate of reduction in the cost of producing hydrogen through electrolysis. On current information, the battery is likely to win out in vehicles used mainly within cities, and hydrogen more likely to be competitive for vehicles travelling over long distances. Hydrogen would be expected to do better for long-distance freight on roads and for rail where traffic is insufficiently dense to support low-cost electrification.

Other countries' investment in research development and commercialisation in competing technologies will define the lower cost paths to electrification. Australian innovation is necessary in developing business models and infrastructure for Australian conditions.

Transport and the grid

The eventual balance between battery and hydrogen routes to electrification will have implications for the shape of the electricity grid. Batteries will be more conducive to decentralised generation and storage of electricity. Economies of scale in electrolysis would drive greater centralisation of energy generation and storage through hydrogen.

Electrification of transport through centralised, hydrogen-based technologies would contribute straightforwardly and considerably to the security and reliability of electricity systems. Decentralised battery-based electrification of road transport could either enhance or detract from grid efficiency, costs and system security and reliability, depending upon power-pricing arrangements. It would be cost-reducing and stabilising if electricity pricing provides incentives to fill batteries when usage of the grid and wholesale prices are low. Otherwise, the battery recharging will increase costs and destabilise the system.

With 100 per cent direct and indirect electrification and current patterns of transport use, Australian electricity demand would rise by about an eighth for passenger cars and a bit more than that again for freight and mass passenger movement. That would increase current levels of power demand by over one-quarter. This would require large increases in renewables generation. For renewables to be supplied from low-cost locations, there would need to be investment in long-distance high-voltage transmission.

The demands on electricity distribution systems would be immensely expensive to service if users of electric vehicles plugged in their vehicles without regard for the economic cost of using grid power at times of peak demand. Marcus Brazil from the University of Melbourne has estimated that with a 10 per cent uptake of electric vehicles, each family's plugging in on arrival home from work between 6 p.m. and 7 p.m. would generate major problems of power system instability.[1]

Large investment in the distribution network would be required to remedy the problem. However, with effective incentives for households to fill their batteries outside the couple of hours of peak demand each day, 80 per cent uptake would be consistent with grid stability without new investment. The associated increased intensity of off-peak grid use would reduce unit costs of the grid by around one-fifth. This would make a substantial contribution to reducing the cost of past wasteful overinvestment in the grid that was identified by the ACCC and discussed in Chapter 4. The more economic use of the grid could go further if electric vehicle batteries were filled at times of low grid demand – say, 11 p.m. to 7 a.m. – and discharged to run the power requirements of the home at peak times – say, 5 p.m. to 10 p.m. There is thus a large economic prize for households and the electricity supply system from integrating the battery-powered electric car with the household power system.

So the shift to electric vehicles could either greatly increase the cost of electricity to all and especially small-scale users, or greatly reduce the cost, depending on whether we have economically suitable power pricing and favourable consumer responses. Until now, Australian regulatory arrangements, far from promoting economically efficient pricing calibrated to the separate wholesale price and grid demand peaks, have provided incentives for businesses in the regulated system to exacerbate peak demand. Increasing peak demand rewards investors in the regulated grid by expanding opportunities for low-risk, profitable investment. The regulatory authorities will need to insist upon time-of-use pricing that shifts the charging of vehicles away from the peaks in grid use and wholesale prices, and encourages discharge from the car battery into the home at time of peak grid demand and into the grid at time of peak wholesale prices. The attitudes and behaviour of consumers are also relevant to designing pricing systems that are effective in shifting demand in ways that reduce costs.[2]

Time-of-use pricing would contribute over time to substantially lower average power costs by reducing the need for investment in the network and in peaking power generation and storage even if we did not have to manage the electrification of transport. The contribution is especially large when vehicle battery-charging is adding to power demand.

Time-of-use pricing would be most effective if it had two elements: one related to times of maximum use of the grid, and one related to wholesale power prices. The latter may be high either because power demand is high or because supply is low at a particular time. The incentives could take the form of rebates for power users who are able to hold total demand below specified levels at peak times as they are defined from time to time.

The pains of adjustment to time-of-use pricing, carefully designed to suit user preferences, would seem to be small compared with the gains. Costs would be lower in Victoria, which already has time-of-use meters.

There are additional household and system benefits from time-of-use-pricing where decentralised solar PV systems play a significant part in power supply. Australian household take-up of solar PV is the highest in the world, and the solar resource richer than in other developed countries. Rooftop solar will continue to expand rapidly as costs fall and with new state-based government assistance, until over half of households are using it. Furthermore, the proportion of households with battery storage is already exceptionally high in Australia and is growing rapidly. As battery costs fall, there will be opportunities for integrating use of car and house batteries in ways that greatly reduce pressure on the grid at peak times. Allowing surplus household battery power storage to be fed back into the grid at peak times could substantially increase the cost advantages of more-efficient grid utilisation that come with time-of-use pricing and the electric car.

Electrification of freight movements through batteries or hydrogen is more easily reconciled with more efficient utilisation of grid capacity.

Large users of power are more responsive to price incentives. They will more quickly and strongly respond to incentives to change patterns of power use with time-of-use pricing.

Demand from passenger rail is inherently concentrated in massive peaks, overlapping with the edges of the solar day. In the Newcastle–Sydney–Wollongong system, peaks of 1 gigawatt are imposed on common loads of about 400 megawatts during daytime, and less at night outside the peaks. Super peaks arise when multiple trains leave stations at the same time in the peak hours. This is expensive electricity demand to serve, with the train company paying high prices. The new energy systems can use fast-response batteries to reduce ordinary peaks and remove super peaks. Efficient use of new battery technologies would ease the trains' pressure on the grid at times of maximum stress.

Australia is a global laggard in taking up battery and hydrogen electric vehicles. In 2018, electric vehicles represented about 0.2 per cent of Australian car sales. This compared with 3.8 per cent in the world and 5.8 per cent in China.

In China, large incentives for purchasing and producing electric cars and public provision of charging facilities have led to rapid expansion of production and sales since 2016. In other countries, recognition of the external costs of local atmospheric pollution and its health effects, as well as of climate change, has led to incentives to accelerate uptake of electric vehicles. As with solar and wind power, the expansion of sales and production in response to incentives has led to reduction in costs. This leads to a virtuous circle of falling costs, rising demand, increased production scale and falling costs. The most effective incentives improve access to charging infrastructure, although consumer and producer subsidies and preferred access to parking and highway space are also important in some countries.

The expansion of production and reduction of costs in China is going to transform the electric vehicle market, as it did earlier for solar PV.

In 2012, China's twelfth Five-Year Plan set a target of 5 million electric vehicles on the road by 2020. This commitment was repeated in the thirteenth (2016–20) plan. China will exceed this target. Total sales exceeded 1 million in 2018 and are estimated to reach 1.6 million in 2019, building on 645,000 units in the first half of the year. This represented 6 per cent of new vehicle sales. In mid-2019 there were over 1 million public electric-vehicle charging stations in China. Many large cities have policies of only using electric buses by the end of 2020. Shenzhen was a leader, with 16,000 electric buses in operation in mid-2019.

The rise of the electric car

Costs come down rapidly at this scale of production. China will lead the world in production of a cheap popular car for mass use. The United States, Germany, Japan and Korea have concentrated on higher quality and more expensive cars, although each has lower-priced models. Tesla, with its high-quality and expensive vehicles, leads the industry in sales from its US base and has recently moved into large-scale production in China.

Meanwhile, global sales and production of cars with internal combustion engines reached a peak in 2017 and have since decreased. Leading car manufacturers are concentrating on research, development and commercialisation of the electric car.

Electric cars are still more expensive than ones with an internal combustion engine. Bloomberg New Energy Finance has brought forward its estimate of the year of capital cost parity between the two types of car several times, and now estimates that it will be 2022.

An electric vehicle has several advantages for the private user. It uses energy more efficiently – around two-thirds to three-fifths less energy for the same distance in a car of similar carrying capacity. Electricity is much cheaper per unit of energy, even when excise is removed from petrol and diesel. There are far fewer moving parts (fewer than eighty) compared to a vehicle with an internal combustion engine (about 1000), so maintenance costs are much lower and longevity greater. Finally, the many amenity advantages of the autonomous car are easier to achieve in an electric vehicle. These factors ensure that the autonomous car of the future will be electric.

Against this can be set the internal combustion engine's advantages at this stage of greater range and more accessible and faster refuelling. These differences are likely to be removed in the future by several developments: in battery and charging technology; use of hydrogen in situations where these differences are crucial; investment in charging infrastructure; new patterns of car ownership and use that allow switching with the purpose of use between hydrogen and battery vehicles; and more intense use of a smaller total number of vehicles to take advantage of the longevity advantage of electric vehicles.

Provision of charging and replenishing infrastructure is the pressing requirement for acceleration of use of electric vehicles in Australia. There is a case for public provision at state or local government level at the early stages of developing new transport systems. Regardless, there will be pressure for all levels of government to invest heavily in public provision of charging facilities, at least while numbers are growing rapidly from their current low base. Developments in the rest of the world will drive the purchase prices of electric vehicles down to and then below that of vehicles with internal combustion engines. From that point, it will not be long before internal combustion engines do not feature much at all in new car sales. Then, with the average life of an internal combustion car in Australia about eighteen years, the number of old-technology cars on the road will decline steadily.

EARTHING CARBON

Australia can make an exceptional contribution to climate action by creating natural systems to store more carbon in soils, pastures, woodland forests and biodiverse plantations. Research a decade ago did not permit definitive assessment of how much carbon could be captured in Australia in these ways. However, in 2011, I speculated that the value of land credits sold into the emissions trading scheme could equal, by 2030, the contribution now made by wool to the Australian farm economy.

My treatment of carbon in the Australian landscape in 2008 and 2011 drew upon pioneering work by the CSIRO and the state departments of agriculture, as well as research at universities. A CSIRO publication in 2011, published after my second Review, highlighted the importance of the opportunity: 'Our soils and forest store large quantities of carbon: somewhere between 100 and 200 times Australia's current annual emissions. We can potentially increase these stores in our rural lands and perhaps store or mitigate enough greenhouse gases to offset up to 20 per cent or more of Australia's emissions during the next 40 years.'[1]

We still can't speak definitively on the size of the opportunity. Australian research funding and effort over the past decade have not matched the economic and environmental importance of the subject.

Meanwhile, what has come to be known as 'natural climate solutions' have become much more prominent in international and especially the European and North American discussion. Recent reports from the Intergovernmental Panel on Climate Change (IPCC) have elevated the importance of capture of carbon in the landscape. It is estimated that natural climate solutions can provide 37 per cent of cost-effective reduction in global carbon emissions for a two-thirds chance of holding warming below 2°C.[2] These reports indicate that native forest restoration and reforestation could sequester up to 480 gigatonnes of carbon dioxide in terrestrial ecosystems – sufficient to meet the negative emissions needs of many 1.5°C scenarios.[3]

The decarbonisation of electricity and the electrification of industry and transport can remove about two-thirds of the reductions to net zero global emissions. The land use, agriculture and food transformation can deliver most of the rest.[4]

A recent research project from the US Academy of Sciences suggested potential for 10 gigatonnes per annum sequestration in global and 1 gigatonne per annum in US landscapes over the period to mid-century during which the world needs to achieve zero net emissions. Australia should have sequestration potential comparable to that of the United States. The low agricultural value of most Australian land reduces the opportunity cost of management for carbon sequestration. It is of national economic consequence that we undertake the research to define the scale of and the means of unlocking the opportunity. In the meantime, the judgement on scale presented in 2011 seems modest.

The big difference now compared with 2011 is that we no longer have the prospect of an emissions trading system into which land-based carbon credits can be sold. Market opportunities are going to be necessary if the potential for Australian carbon farming is to be realised. Australia can nevertheless make a start with current policies and go further in ways that do not violate the explicit electoral commitments of the current federal government. That will provide time for this or successor governments to craft new policies that allow the opportunities for rural Australia to come to fruition.

Compared to other nations, Australia has two advantages in capturing carbon in the landscape. The first is our exceptionally large endowment of woodlands, forests and other land relative to population. The second is our exceptional expertise in land-based industries – from agricultural and forestry science, through agricultural and resource economics to public and private knowledge and institutional arrangements supporting commercial success. Advanced knowledge and innovation were necessary for transplanting European-style agriculture to a strange and unpropitious physical environment. Research, innovation and education supported by public institutions were important from the earliest times. Visiting in the late nineteenth century, Mark Twain gave agricultural research in Horsham in western Victoria an honourable mention. The CSIRO had its origins in applied research for the agricultural and pastoral industries. The distinctively Japanese system for trading rural (and, later, all) commodities through multi-product, integrated trading companies had its origins in the Japan–Australia wool trade. Wool futures were traded in Sydney before the establishment of the Chicago Mercantile Exchange. Australia was centrally important in developing the international agricultural research system and continues to contribute a great deal to it.

A decade ago, the international rules for measuring land use emissions were immature, and had to be developed on sound principles to make the most of global opportunities, and to recognise Australian contributions. Other developed countries that had shaped the system did not share our opportunity or interest. Australian expertise in developing environmentally and economically efficient rules on land use could be of particular assistance to Australia's neighbours in southeast Asia and the south Pacific.

The special Australian place in absorption of carbon in the landscape was the subject of remarkable commentary eight decades ago by the first Distinguished Fellow of the Economic Society of Australia, Colin Clark. In his 1940 book *Conditions of Economic Progress,* Clark helped to found development economics and the use of national accounts for production and income. Clark sought to allay concerns that the dependence of modern economic development on fossil energy and the finite nature of coal and oil resources would bring economic growth to an end. He noted that we can calculate the likely amount of undiscovered fossil fuels from the carbon that was once in the atmosphere. 'However, we must not set out to burn them up too fast, even if we do find them, at any rate not faster than the rate at which the carbon dioxide can be converted by photosynthesis.'[5] However, he reassured us that keeping the use of fossil fuels within the limits of what can be absorbed by photosynthesis need not be the end of economic growth. An abundance of solar energy falls on the earth, we just need to know how to tap it. The best method at present, he said, is the proven process of photosynthesis in trees, and he calculated that the eucalypt is the most productive agent for conversion of solar energy into biomass at present. Algae had the potential to do better. The silicon battery and other recent discoveries, he said, may do better still some day.[6]

So the importance of the Australian eucalypt to sustaining economic growth without excessive carbonisation of the atmosphere was recognised eighty years ago, at the beginning of development economics.

My own first close professional contact with climate change came through rural development. I was chair of the International Food Policy Research Institute in Washington in the 2000s, when analysis of future food security identified climate change as a long-term threat in south Asia and Africa in particular. The first decade of the new century had

① SEE P142

seen an historic lift in world food prices. I noted that rising living standards and increasing demand for high-quality food in developing countries would make this an expanding opportunity over a long period for Australian farmers – unless climate change at home damaged Australian supply capacity. (1)

In 2008, I brought into the mainstream discussion some early work by the CSIRO and state departments of agriculture on the immense mitigation potential of changes in land use. Nurturing vegetation on the dry, degraded mulga country where rainfall was spasmodic in Queensland and New South Wales could be transformative. Innovative uses of the properties of Australian eucalypts included farming of the mallee on the arid boundaries of crop cultivation for subterranean sequestration and for harvesting biomass.

The 2011 Review took the land use mitigation story further. It advocated inclusion of offsets from agriculture into the emissions trading scheme through what became the Carbon Farming Initiative (CFI). These arrangements were carried into the Abbott, Turnbull and Morrison governments' Emissions Reduction Fund (ERF). The ERF was a clunky, truncated and less adequately funded version of the CFI. It required resources from general revenue, rather than from sales of emissions permits. Nevertheless, Abbott's ERF kept alive the sale of offsets as a way of providing incentives for farm sequestration. The arrangements developed by the Clean Energy Regulator showed how an offsets scheme related to land use could work, and that there was strong private response to incentives.

In 2011, I also sought to place climate change into a broader story of global development. The rise in living standards and fall in fertility in the early modern Western Europe had allowed part – and over time an increasing proportion – of humanity to free itself from the Malthusian trap that had held back the living standards of ordinary people since the beginnings of agriculture about ten millennia ago. Failure to mitigate climate change would lock many people in poor developing countries today into the Malthusian trap, as declining food production fed back into living standards and fertility.

What progress has been made since 2011 on Australia's opportunities in the landscape? The positive outcomes are substantial, but we do not yet understand the extent of the potential. The unusually large endowment of land and woodlands relative to population gives Australia immense advantages in the production of biomass, as well as in the capture of carbon in the landscape.

Over the past eight years there has been increasing global awareness that transformation in the landscape and of food and raw material production and consumption is necessary if the high material standards of living enjoyed in the developed countries are to be made available to all of humanity. If the high consumption per person of land-intensive meat produced in traditional ways were to spread from developed countries to the whole of humanity, large and disruptive increases in food prices would ensue. We saw the beginnings of that as rising Chinese incomes and consumption put pressure on global markets through the first decade of this century. Transformation is also necessary for improved human nutrition and health – not to speak of wider ecological sustainainability. And it turns out that the transformation in agriculture and land use is also necessary to contain the cost and effect of climate change.

Recent research has tracked the increase in grain prices as people in developing countries consume more meat as their incomes grow. The EAT-Lancet Commission Report on Food, Planet, Health, released early in 2019, showed that a large fall in global average consumption of red meats and an increase in consumption of legumes, pulses and other plant-based foods would improve human health on a global scale.[7] And later that year, the report from the Intergovernmental Science-Policy Platform on Biodiversity and Ecosystem Services drew attention to a biodiversity crisis that threatens ecological systems of fundamental importance to humanity.[8]

Expansion of knowledge in all of these ways points to the need to reduce the size and weight of the human footprint on the planet: to consume less land- and emissions-intensive food, including through reducing per capita consumption of red meats in high-income countries; to increase forested tree-cover and its diversity; to economise on use of water; and to do this while drawing on renewable sources of the chemicals for industrial processes that were previously supplied by coal, oil and gas.

At the same time, knowledge has been expanding about how to achieve this: knowledge of nutritious and palatable substitutes for old foods that have less damaging ecological impacts; knowledge of land and farm management systems and their long-term viability and ways to improve the relationship between environmental cost and value in use; and knowledge of regulatory arrangements that can encourage the necessary transformation.

There is rising awareness of the need for fundamental change, and the beginnings of a sense that such change is feasible. In this, 2019 for landscape feels like 2011 for the transition from fossil to renewable energy. The anticipated costs of change are still high, but increases in the scale of the new are starting to bring their costs down. Transformation is unfamiliar, and daunting, but beginning.

Australia is uniquely well placed to lead and prosper from the land use transformation, just as it is in energy and energy-intensive manufacturing.

Landscape and emissions

Land-use change and agriculture were unusually large in Australian emissions in 1990 and 2008, exceeded proportionately only by Ireland and New Zealand among the developed countries.

① SEE P 142

At first glance, change in forest and woodland emissions is the great success story of Australian mitigation. Such emissions fell dramatically through the 1990s with restrictions in land clearing, especially in Queensland (Chart 7.1). They then stabilised for over a decade from 1996. They resumed their decline from 2007. They were negative from 2012 – the only Australian sector with negative emissions then or in prospect. Since then, they have stabilised around that negative level. This reduction offset a massive blowout in emissions from other sources.

Chart 7.2 shows the three main contributors to this turnaround: the decline in landclearing; the growth of old forests; and new conversion of pastoral and agricultural land to plantations. The reduction in landclearing was mainly the result of tighter regulation in the first half of the 1990s and again from 2007 to 2013, followed by some loosening after 2013. The expansion of biomass in plantations was strongly influenced by taxation incentives introduced in 1997 and made less attractive by withdrawal of concessions from 2006.

For agriculture, Australian emissions were exceptionally high, mainly from enteric emissions from cattle and sheep. A small decline in agricultural emissions between 1990 and 2008 was chiefly the result of low rainfall reducing sheep and cattle numbers (Chart 7.3). The fall ceased and slightly reversed after 2008, with better seasonal conditions.

Australia's unusually large endowment per person of woodlands and forests (Chart 7.4) confers massive advantages for sequestration and for production of biomass for industrial use.

① Sequestration potential is affected by rainfall. The applied atmospheric physics is still coming to grips with the effects of warming on rainfall in particular locations. In the world as a whole, average rainfall rises, but the distribution of where it falls changes. Scientific advice says that southern Australian latitudes are likely to experience lower rainfall, and, on the whole, northern latitudes higher.

Chart 7.1 Land use emissions 1990 to 2018

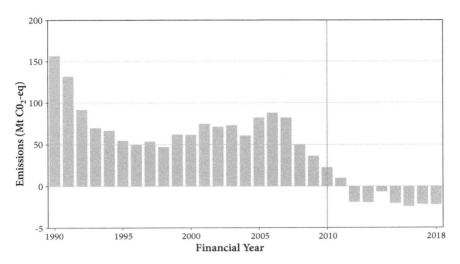

Note: The data are reported by financial year.
Source: Department of the Environment and Energy, accessed 5 May 2019.

Chart 7.2 Land use emissions 2000, 2010, 2018

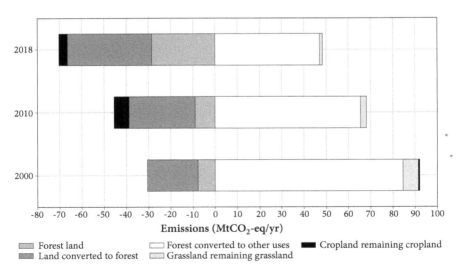

Note: Grassland remaining grassland includes Wetlands and Settlements.
Source: Department of the Environment and Energy, accessed 5 May 2019.

Chart 7.3 Agriculture emissions 2000, 2010, 2018

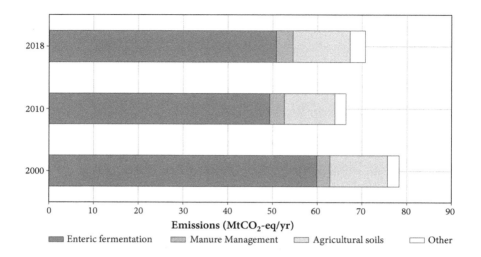

Note: Other includes rice cultivation, burning of agricultural residues and savannahs and liming and urea application.
Source: Department of the Environment and Energy, accessed 5 May 2019.

The southern latitudes have long been the source of most Australian agricultural value. With agriculture challenged by the combination of drying and warming, there will be shrinkage in the area of cultivation. Newly submarginal farmland may be more productive as a source of carbon sequestration and biomass. Wetter northern areas will be able to sustain more intense living plant concentrations, making them potentially larger sources of biomass for harvesting or long-term carbon sequestration.

The 2008 Review took a partial look at the opportunities. It suggested that new patterns of management of semi-arid rangelands had the potential to sequester around 250 million tonnes per annum of carbon dioxide equivalent; native forest regeneration perhaps 140 million tonnes; plantations in high rainfall areas approaching 140 million tonnes per annum; and smaller numbers in a range of other activities.

Chart 7.4 Per capita area of forest and wooded land 2015

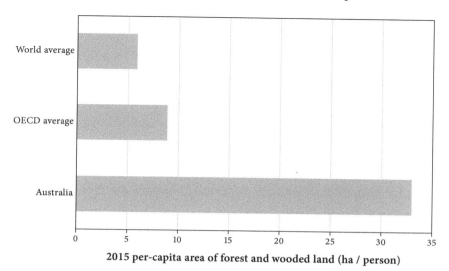

2015 per-capita area of forest and wooded land (ha / person)

Note: The FAO category used for data is 'tree-covered land', and 2015 is the most recent year of data availability.
Source: FAO 2015, accessed 5 May 2018.

No estimate was provided for the sequestration potential of soils, but Australian micro-studies and larger projects in the US and Europe suggest that this could be as large as the potential in woodlands and forests. Partial studies since then have focused on subsets of these opportunities.

The clear point to emerge is that the potential is immense, and that large investment in research is warranted. Use of the potential requires sensitivity to interactions across a complex ecology, including with water balances and biodiversity. My own assessment is that good management and provision of efficient incentives for sequestration in soils and the range of vegetation would support sequestration of carbon of at least hundreds of millions of tonnes per annum. In a globally efficient sequestration effort, this would see Australia emerge as a major exporter of carbon credits from the management of land and vegetation, alongside rapid reduction of domestic emissions from combustion of fossil carbon.

The potential for forest and woodland sequestration in the extensive pastoral regions is poorly understood, but immense. Arid and semi-arid rangelands make up about 70 per cent of Australia's landmass, or around 5.5 million square kilometres (550 million hectares). This is mostly grazed by cattle and sheep currently contributing little economic value. Early work by the CSIRO, suggesting large sequestration potential, focused on restoration of vegetation by removal of low-value pastoral activity in the mulga country. These estimates depended on small carbon increases per hectare per annum across very large areas of land. One Queensland study estimated that rehabilitating 200 million hectares of overgrazed rangelands could sequester 100 million tonnes per year over forty years, with one-fifth coming from restoration of degraded mulga country.[9] Witt et al. (2011) pointed to 1.1 tonnes per hectare after destocking. The same study suggested 25 to 30 metric tonnes per annum potential in the mulga area of Queensland. A later Queensland government study suggested that reduced grazing pressure without complete removal can still lead to substantial carbon gains in mulga country.[10]

The Queensland and New South Wales mulga country is only part of the pastoral country in which the accumulation of carbon in trees could be of major importance. To maximise value of production from this land across all possible uses would require a reward for carbon sequestration and separate fiscal incentives for biodiversity, leaving landowners to judge how much of their land should be subject to destocking for carbon and biodiversity purposes.

The mallee is a eucalypt marvellously adapted to the arid, fire-prone Australian environment. Its trunk grows beneath the land surface, where it is not affected by fire. A bushfire can consume the multiple branches with leaves sprouting from the trunk below the surface, and they will quickly grow again. This allows regular harvesting for processing of large tonnages every four years – around 8 tonnes per annum per hectare in the dry West Australian wheat belt. Sequestration continues through increases in the mass of the trunk (the 'mallee root') while the branches and leaves provide a periodic harvest of biomass.

Mallee carbon development can provide an alternative land use of considerable value in parts of the West Australian, South Australian, northwestern Victoria and southern New South Wales wheat belts that are affected adversely by warming and drying. It also has potential with lower annual yields in areas adjacent to and beyond the old wheat belt boundaries.

Research and development at Curtin University funded partly by ARENA has developed technology for decentralised conversion of bulky biomass into concentrated carbon and liquid hydrocarbon. A variation on this theme has been applied commercially by a Melbourne-based company. Two valuable commodities are produced. One is a char, or charcoal – a renewable carbon that can be used as a replacement for coal and gas in many industrial activities, or as a source of soil carbon. The second is a gaseous or liquid hydrocarbon, which can be used as a replacement for coal, oil and gas as an industrial feedstock, in power generation or, after refining, transport. The high value relative to volume and mass of these products allows truck or rail transport to major industrial centres or ports at relatively low cost.

Mulga in northern New South Wales through to Central Queensland, like the mallee, rejuvenates after fire or harvesting and could be used as a source of biomass and biosequestration.

Savannah covers about one-fifth of the earth's surface, including much of northern Australia. Climate change is tending to increase rainfall in the Australian savannah country. This is supporting denser vegetation. This is happening naturally, and there are opportunities for human intervention to increase the carbon stock.

The main intervention so far has been timely burning by Indigenous groups to generate new growth and prevent wild hot fires later in the season. Recent studies suggest about 7 million tonnes per annum potential. The Western Arnhem Land Fire Abatement (WALFA) project made the world's first use of financial incentives to increase the carbon stock through timely savannah burning. It has since been repeated across northern Australia, with seventy-five projects having been registered with the Emissions Reduction Fund by January 2018. These had committed to abate 14 million tonnes over eight years. Australian technology based on Indigenous practices could be taken to similar opportunities elsewhere in the world. It has been applied in drier areas of Papua New Guinea in the vicinity of the lower Fly River.

Globally, there is more carbon in soils down to 2 metres than in the atmosphere and living biomass combined.[11] Recent estimates point to very large opportunities for sequestration in Australian soils at relatively low and sometimes zero or negative costs.[12] There is greater awareness now of the loss of organic soil carbon and the risk of more with warming and drying in southern Australia[13] – although this must be matched against opposite tendencies in the wetter northern latitudes experiencing greater rainfall. There is considerable potential for use of biochar from renewable sources to increase soil carbon.

The application of charcoal or char to agricultural land to magnify increase in carbon has soil productivity as well as sequestration benefits. Increase in soil carbon is one element of increasing global interest in regenerative farming, in which Australian landowners have been prominent.[14] These landowners capture more solar energy and develop biodiversity in vegetation, restore soil biota, capture and store rainfall in soil and slow water flows, plant trees on a significant proportion of their land and increase economic profitability with more realistic expectations of long-term sustainability. Much learning in Australia came from seeking to understand the pre-European landscape as well as directly from Indigenous residents and storytellers. 'Regenerative farmers' take note of each element of the landscape as an essential component of the whole. Costly input use is diminished or discarded as natural resources drive their ecosystems. Soil organic matter is retained from decaying and composted vegetative material. Grazing animals become means of spurring new growth and cycling and distributing nutrients through dung and urine. Mechanical soil disturbance is minimised. Organic and inorganic soil carbon increases. More phorporous and other valuable minerals cycle through the bio-system. Biodiversity of birds and insects increases. Disease in plants and animals falls. Overall nutrient value of food for human consumption increases.

Methodologies for providing credits for increases in soil carbon have been recognised in the Carbon Farming Initiative and the Emissions Reduction Fund. Transaction costs have been high, largely due to high cost of measurement. Reducing this cost is the prime focus of the recent interactions between University of Melbourne researchers and former governor general Michael Jeffrey, founder of the regenerative farming advocacy group Soils for Life. Jeffrey's interest has recently been supported publicly by Prime Minister Scott Morrison.

The general story is of immense potential for sequestration of carbon through changes in Australian landscapes, but of small and diminishing research effort to define the potential and the means of unlocking it. Two developments have contributed to this unsatisfactory situation: the absence of generally available incentives and a general reduction in research and development on agriculture, pastoral activities, forestry and climate change.

A major proportion of Australian emissions in agriculture are from enteric fermentation in cattle and sheep. Opportunities for reducing emissions include biological interventions, changes in feed regimes, selective breeding and shifts in grazing patterns. Recent research supports the possibility of saving about 16 million tonnes per annum, or more, from such changes in stock management. When we again have incentives to reduce methane emissions, extending to the agricultural sector as in New Zealand, there is potential from additives to livestock feed.[15]

More efficient farm and station management generally can reduce emissions per unit of output by large amounts. It generates more value from the same number of animals, or the same value from smaller numbers.

The most powerful influence on total enteric emissions is cattle and sheep numbers. This will be governed by the profitability of various land uses. The introduction of a carbon price and incentives for carbon sequestration on extensive pastoral land would reduce cattle and sheep numbers on marginal country with relatively little loss of economic value in pastoral activities.

One major study has investigated harvesting kangaroos in place of ruminants in some Australian locations.[16] Large-scale shooting of kangaroos is conducted under licence to control numbers. Most of the carcasses rot where they fall. Only 3 per cent of the sustainable kangaroo yield in the wild environment is currently harvested for human use. There is unexplored potential for more systematic harvesting of kangaroos from pastoral regions as well as for reduction of waste. Kangaroo meat has health as well as mitigation advantages, and marsupials fit more easily than cattle or sheep into general efforts to restore vegetation and biodiversity.

There is considerable potential to reduce emissions through use of agricultural, woodland and forest output for bioenergy production. Farine et al. (2012) have suggested that first- and second-generation bioenergy (crop residues, forest product waste) could remove 26 million tonnes per annum of emissions.[17] Thirteen million tonnes per annum of emissions could be avoided through use of second-generation bioenergy production systems (for example, oil from algae and the oilseed tree *Pongamia pinnata*). The same authors have suggested that ligno-cellulosic biomass production from the mallee could replace 48 million tonnes per annum of emissions (28 per cent of emissions from electricity generation or 9 per cent of total emissions).

In the zero-carbon economy of the future, however, biomass is likely to be reserved by its price for high-value uses in civil aviation and inputs into chemical industrial processes. The value would be large, but it may not contribute directly to reducing Australian emissions as it or products made from it are exported and replace emissions from fossil energy throughout the world.

The food challenge

The twenty-first century has seen strong economic growth in developing countries, even as incomes have stagnated in the developed world since the global financial crisis. Rising incomes in developing countries have seen rapid movement towards the food consumption patterns of the rich. Chinese meat consumption rose more than tenfold over forty years of reform and development, to almost 70 kilograms per capita per annum, or about the average of the developed countries. Chinese consumption is dominated by non-ruminant meats – pork and chicken – which are more efficient in converting grain into animal protein and do not emit methane from enteric fermentation, but nevertheless place pressure on land and the natural environment. Chinese direct and indirect demand for grain has pushed up global prices.

South Asia, with its religious and cultural constraints on eating meat (particularly beef), will not see similar levels of meat consumption to China. But the pressures on land resources from increasing expenditure on food will still be large. And development in Africa is likely to place similar pressures to those in China on food production and the natural environment.

The extension of something like the developed world's pattern of food consumption to the whole of humanity is inconsistent with climate stability, and more generally with the stability of ecological systems. It would also be unhealthy.

The convergence of economic development, climate, biodiversity and general ecological and health imperatives will transform global diets. Rising prices of meat in general and especially ruminant meat will modify consumption patterns. We think of our food consumption patterns as changing little over time. In fact, the experience is that Australian food consumption has changed radically in response to exposure to alternatives through culturally diverse immigration, and to such mundane influences as relative prices. Australians led the world in consumption of sheep and cattle meat when I was young. Since 1960, Australian consumption per person of meat from cattle has fallen by almost 20 per cent, and from sheep by almost 90 per cent, despite very large increases in average incomes. Over the same period, pork consumption has increased by over 375 per cent and poultry by over 1300 per cent. Consumption changed inversely to relative prices: between 1960 and 2018, poultry real prices fell by 75 per cent, and pork rose only a little (having fallen well below 1960 levels through the 1980s, 1990s and 2000s), while beef prices rose by over a third and mutton and lamb by over two-thirds. Even in a conservative country like Australia, food consumption patterns can change radically over relatively short periods of time. And it seems that changes in relative prices are an important force for change.

Changing personal preferences will play an important role in the future of food. The health story will become more influential everywhere – following trends among better-educated and higher-income people in the developed countries. Larger numbers of people in high-income nations will be influenced by the climate, ecological and animal welfare cases for greater reliance on plant-based foods.

Changing personal preferences, relative prices and technological opportunities will open the way to a revolution in production of meat-like manufactured foods. Products from processes using animal stem cells to grow chemically indistinguishable meat substitutes from biomass will also become indistinguishable in taste, texture and nutrition to meat from the killed animal – and superior in nutrition, if that is preferred. Vegetable-based foods with the taste and texture of meat will become closer to the real product for those who prefer it to unaltered natural vegetarian fare. The price of meat from both stem cells and plants will fall, eventually to below the rising price of farm meats. The first cell-based hamburger a few years ago cost $US300,000 to produce. In 2019, it could be produced for about $US10. Large-scale production will bring costs down further – to lower than farm-based meats of all kinds and far lower than the meat of cattle and sheep.

More generally, there will be a shift in demand to higher-quality, safer and more expensive food. Australia has a comparative advantage in supply of the higher-value products that will dominate demand – especially in supplying the world's largest and most rapidly growing markets in Asia. The main meat substitutes make intensive use of biomass and energy and use less water than the animal competitors.

The new foods will include growing proportions from hydroponic technologies using renewable energy. The Sundrop greenhouse in Port Augusta, producing tomatoes to supply Coles retail outlets all over the country from desalinated water using solar energy in an arid region, is an exemplar of a new order. Capital and expertise will be more important and water less so in food production.

Food production in the old southern Australian centres of agriculture will be challenged by rising temperatures and falling rainfall and run-off into rivers. Water will be increasingly scarce and valuable. Many of the agricultural regions suffering from reduced water availability have access to low-cost renewable energy. Together with technological development in pumping, desalinating, treating wastes, raising the quality and more efficient use of water, this will provide opportunities for increasing the availability of water for agriculture and reducing the damage from climate change.

Carbon farming and the international rules

In 2008, I concluded that climate mitigation in the land sector required comprehensive carbon accounting. The inclusion of land under the Kyoto Protocol framework was incomplete. With the adoption of the Paris Agreement in 2015, and the subsequent rulebook adopted at the end of 2018, all countries will be required to report emissions under the same United Nations Framework Convention on Climate Change (UNFCCC) reporting framework, applying the latest guidance from the Intergovernmental Panel on Climate Change (IPCC), which includes a more comprehensive approach to land-based accounting.

The Paris Agreement raises an expectation that the long-term mitigation goal will be achieved through a 'balance between anthropogenic emissions by sources and removals by sinks'.[18] This highlights terrestrial sequestration of carbon emissions.[19]

While reliance on land-based mitigation in the Kyoto Protocol was subject to strict limitations and caps, in the Paris Agreement land-based sinks are now prominent.[20] A reporting, review and stocktake architecture was agreed in Warsaw to implement the Paris Agreement.

One weakness of the Kyoto Protocol pointed out by the 2008 Review was that harvested wood products (HWPs) were not included in the first commitment period. Instantaneous oxidation on harvest was assumed. HWPs are now included, but the inclusion is complicated by the new rules. For deforestation, grassland or cropland, an instantaneous oxidation approach is used, which effectively means HWPs are not counted.

An activity-based approach still provides the primary architecture for accounting for land-use change and forestry. A move to more comprehensive accounting is needed to fully capture terrestrial carbon fluxes. In practice, as the activity approach becomes more comprehensive, the results tend to approximate those of the land-based approach.[21]

Australia's nationally determined contribution (NDC) under the Paris Agreement is an economy-wide target, meaning it covers all sectors of the economy, including agriculture and land use, land-use change and forestry.

Australia has already taken a step towards more comprehensive accounting (with wetlands being the only Kyoto Protocol activity it is not accounting for) and is well positioned to fully account for land-sector emissions and removals under the Paris Agreement.

The Carbon Farming Initiative (CFI) allowed farmers and land managers to earn Australian Carbon Credit Units (ACCUs). Each ACCU represents one tonne of carbon dioxide equivalent stored or avoided by reducing greenhouse-gas emissions. The ACCUs could be sold to clear obligations under the carbon-pricing rules. In July 2014, the carbon price was repealed. On 31 October 2014, the new Coalition government's climate strategy, the Direct Action Plan, was passed, which established the Emissions Reduction Fund (ERF).[22] The shift was made from a carbon price to government-purchased abatement, and an expanded CFI, moving eligible projects beyond the land sector to include energy and transport. In the ERF, $2.55 billion was made available for direct purchasing of abatement under the reverse auctions, of which $226 million remained in May 2019. The government's Climate Solutions Fund was announced on 25 February 2019 to appropriate an additional $2 billion from 2020–21 onwards to fund auctions to 2030.

The ERF involves a voluntary crediting and purchasing mechanism. To ensure these emissions reductions are not displaced significantly by a rise in emissions elsewhere in the economy, a safeguard mechanism requires Australia's largest emitters to keep net emissions below baseline (historical) levels. The safeguard mechanism applies to around 140 businesses that have direct emissions of more than 100,000 tonnes of carbon dioxide equivalent a year.

Projects that meet the requirements under the various methodologies can generate ACCUs for emissions reductions. Projects can sell their ACCUs on the voluntary market, or bid to sell them to the government in auctions run by the Clean Energy Regulator. Auctions are held twice a year. The ninth ERF auction was held on 24–25 July 2019. The average price per ACCU contracted has been $11.92 over the life of the scheme, with the average price at individual auctions ranging between $10.23 (April 2016) and $13.95 (April 2015).

The fund provides a broad range of opportunities to reduce emissions across the economy. Over 750 projects have been registered, including improving energy efficiency in buildings and industrial facilities, diverting waste from landfill and using gas from wastewater to generate electricity, and reducing emissions on the land by protecting native forest that would otherwise have been cleared, savannah management, reducing emissions from beef cattle production, or revegetating marginal country. Vegetation methodologies make up the greatest number of projects. These activities are dominated by New South Wales and Queensland. The greatest activity has been where the project costs are likely to be lowest (avoided landclearing and allowing cleared land to regenerate). Low uptake in Victoria may be due to higher land prices.

Agriculture and land use were spared the general removal of incentives for reducing carbon emissions under the Abbott–Turnbull–Morrison government by the establishment of the ERF. This kept alive important features of the CFI, including its administrative framework. The loss of opportunity for selling offsets into an emissions trading scheme lowered the price to well below the social cost of carbon, placed a low cap on total funding and, by making access to credits dependent on winning a bid at auction, introduced uncertainty about whether action to reduce emissions would be rewarded at all.

These were all setbacks for Australia getting good value from its immense potential for storing carbon in the landscape. There is a way forward that does not violate the current government's electoral commitments. The first step would be to make the whole of the funding for the Climate Solutions Funds available for use now as legitimate carbon credits are certified by the Clean Energy Regulator. This would see the new fund exhausted over a few years.

The second step would be to require in the next parliamentary term, with the necessary electoral preparation, the beginning of phasing in of full offsetting of fugitive emissions by purchase of ACCUs. The full offsetting would be completed through the 2020s. Demand for credits from the farm sector would be further enhanced by the current requirement for all exceedance of baseline emissions within the Abbott safeguard mechanism to be accompanied by surrender to the Clean Energy Regulator of ACCUs. Alternatively, state governments through their mineral leasing or environmental powers could require offsetting of fugitive emissions by use of certified ACCUs – sourced from their own territory, as the local politics would favour expansion of opportunity for the local farm and station community. This is the approach proposed by the WA Environmental Protection Agency in 2019, in its Greenhouse Gas Assessment Guidelines.

Further into the future, when Australia's international climate change mitigation credentials have been restored, linking to the European Union emissions trading system would avoid truncation of the mitigation effort. Time would be needed to negotiate change in European and Australian rules on trade in carbon credits. There would be initial European scepticism about the legitimacy of a number of Australia's rules on farm credits. Where warranted, adjustments could be made. At the same time, Australia would need to persuade European policy-makers of the value of soundly measured and administered carbon farming. Our efforts in persuasion would be supported by the recognition growing in the international community, including in recent IPCC reports noting the importance of natural climate solutions to the global mitigation effort.

Growing the future

It is now clear to the international community – as it was not eleven or eight years ago – that changes in land use and agriculture will have a central role in avoiding high costs of climate change. If we move too slowly and overshoot the Paris targets, soil- and plant-based sequestration – including through the capture of carbon wastes from plant-based industrial processes and storing or using them in ways that keep them out of the atmosphere – will be the main avenue to achieve negative emissions.

The transformation of food, agriculture and land use that is necessary for climate change mitigation is also needed for global development, to improve human health and to maintain a stable global ecology more generally. There will be one agricultural and land use transformation to serve these four great purposes.

To make good use of this opportunity, Australia will need systematic incentives for reducing emissions in agriculture and land, and to provide sound reasons to believe that they are here to stay. And it will need to restore old national strengths that have been allowed to decline in recent years: our strengths in research and education on agricultural, pastoral, forestry and related industrial activities. The combination of low-cost renewable energy and abundant land for biomass will be powerful in the synthetic food production industries of a zero-carbon world.

Alongside our industrial opportunity in renewable energy, our strength in growing and using biomass will set Australia up as the superpower of the low-carbon world economy.

EMBRACING AUSTRALIA'S LOW-CARBON OPPORTUNITY

There is a chasm between a world that quickly breaks the link between modern economic growth and carbon emissions, and a world that fails to do so. The side of the chasm that we are now on is a dangerous place. It would be reckless beyond the normal human irrationality for us to stay where we are.

Australian prosperity and security, as well as our natural and human heritage, will be challenged in fundamental and perhaps unanswerable ways if humanity does not succeed in holding temperature increases below 2°C and as close as possible to 1.5°C. With only half the warming we can expect from 1.5°C, we have already had to deal with dreadful impacts of more severe, earlier and more frequent bushfires; of reduced flows into the Murray–Darling river system; degradation of the Great Barrier Reef; a shift to desalination to supply water for Perth; reduced moisture in our southern farming soils; and high tides lapping at the steps of the beach huts at Brighton in Victoria.

We need to build the bridge on which Australians can walk over that chasm, from policy incoherence to hope and opportunity. Hope that we might avoid the worst outcomes of climate change. Opportunity for Australia to be the world's main trading source of metals, other energy-intensive goods and carbon chemical manufactures in tomorrow's zero-net-emissions world; and a major contributor to the world's efforts to absorb excessive carbon into land and plants.

We need that bridge. And we need Australians to walk across it.

It has always been true that Australians walking across the bridge alone will not deal with the challenge. A large part of humanity needs to make the shift. But many others are already halfway across their bridges and we remain on the side of troubles, encouraging others to stay with us.

We know that we could make the move, fully and decisively, and yet still be destroyed by others remaining on the side of troubles. China, most important of all to this story, was moving across, and has stalled while it deals with its trade war across the Pacific. The United States, second-most important in global emissions and still the leader of international political trends, was striding out along the bridge until a new president in 2017 set out to turn his country back. He has not yet succeeded, but may do so if given long enough at the task.

And beyond international politics, the scientific uncertainties are such that we could meet current expectations of what is required to hold temperature increases to 1.5°C and still have to deal with immense harmful effects.

Australia has the strongest interest among developed countries in the success of a global effort on climate change. Even if we abstract from the effects of climate change itself, we have the most to gain economically from being part of a global transition to a zero-emissions economy. But we are stuck on the side of the chasm with the people who are against effective action.

Does it matter that we are on the wrong side of the chasm – a force against, rather than for, effective action? Won't the important decisions be made in other countries?

It matters morally, for reasons explained by Pope Francis and Professor John Broome. It matters economically: we deny ourselves the vast opportunities for expanding Australian employment and incomes, which are located disproportionately in rural and provincial Australia. These gains would be much larger if we joined the whole world in taking strong action on reducing emissions, but they are substantial even with the world in its current state.

Australian steps also affect the prospect of others building and moving across their own bridges – never as a single determinant of the decisions of another, but as one among many significant influences. Whether Australia actively supports international action on climate change or continues as a drag on the transition will be influential at the margins of decisions in many countries – influential at the margins in ways that sometimes make the difference. It seems crazy now and will seem crazier to Australians in the future if we use the influence we have against rather than in favour of the prosperity and security and natural and human heritage of Australians who come after us.

I am often asked if I am optimistic or pessimistic about Australia and the world doing what must be done to avoid great disruption from climate change. People who look on and despair at past and present failures of action ask how I manage to carry on trying to improve the outcomes or at least the chances of a less damaging outcome.

I reply that there is still a chance of avoiding disastrous outcomes, and the incidental advantages of that outcome for Australians are so large that once my fellow citizens see them as they are, they will want us all to cross the bridge.

Of course, that does not deny the awful reality of where we have gone over the past decade, and where we are today. But Australians with good leadership have made big changes in the past when severe challenges required them.

Neither does it deny the difficulties that we have brought upon ourselves and will leave to future Australians by leaving the start so late and closing off the less tangled paths to the bridge across the chasm. The awful reality is that we may fail to change far and fast enough, and that our grandchildren will inherit a parched and disordered country in which a past time of prosperity, democracy and good order is a myth of origin. Yes, that is a possibility. And yes, sometimes I do think that the time of hope has passed. But there is still a path to a manageable outcome. And I see no good purpose in acceding to despair while the path to a manageable outcome remains open to us.

A decade ago, Australia was deeply invested in the old energy economy, with the highest carbon emissions per person in the developed world. Adjustment to the low-carbon economy of the future was necessary, but expensive. A large part of the cost of doing our fair share in a global mitigation effort would be the loss of coal and gas exports. We would take these losses from other countries' action on climate change, even if Australia did nothing itself. But Australian benefits would exceed the costs if we did our fair share in a global effort to hold human-induced temperature increases to 3°C. We would avoid immense damage to the economy and wellbeing of Australians who come after us at a significant but manageable early cost. Benefits would exceed costs by a larger margin for 2°C.

In 2019, Australia is more deeply invested than ever in the old energy economy. But now the immediate costs of making the change are much lower, and it is clearer that Australia can prosper exceptionally in the post-carbon world. Industries built around removing global emissions are better fuelled from Australia than anywhere else.

In Chapter 2, I summarised developments in the atmospheric physics of climate change, and in the global and Australian struggles to respond to it. I described the battle everywhere between the diabolical policy problem and the saving grace; between the public interest and the vested private interest. The saving grace was widespread community interest in the problem and engagement in finding ways to solve it. Globally, that saving grace has been gradually getting the upper hand, but in Australia the signs are not so clear.

I noted that three electoral events in the second half of 2018 pointed to a retreat of the diabolical in Australia, bringing us back towards the global mainstream. The by-elections in the House of Representatives seats of Longman in Brisbane and Wentworth in Sydney favoured candidates who argued for strong action on climate change. And the Victorian state election returned with an increased majority a government that made promotion of renewable energy and reduction of greenhouse gases a central concern. In counterpoint, the federal election result in May 2019 throws these conclusions into question.

In Chapter 2, I used exit polls from the 2013 election to show that carbon pricing was not an electoral negative of the Labor government. Climate change certainly was decisive in the competition for leadership within the Coalition parties, but that is a different matter.

We have no comparable exit polling for the 2019 election. The strong result for the Coalition government, which had much lower targets and generally weaker policies for reducing carbon dioxide emissions than the Labor Opposition, has been interpreted as electoral support for that stance. Those who interpret the result this way point to the exceptionally strong support for the Coalition in the coal-exporting regions – the Hunter Valley in New South Wales, and the coastal cities and close inland from Gladstone to Townsville in Queensland. But it would be a mistake, and for the government a dangerous one, to conclude that the electorate as a whole was opposed or even indifferent to strong action to reduce greenhouse-gas emissions. The voters in some electorates in southern Australia favoured independent candidates who made strong climate action the basis of their campaigns, most dramatically in former prime minister Tony Abbott's seat of Warringah in Sydney.

The Opposition suffered from a new coalmine in Central Queensland becoming emblematic of the climate change issue. It was not the real climate issue. The international community agreed at the United

Nations conference at Kyoto in 1997 and has maintained since then that each country's contribution to the global mitigation effort would be measured by emissions inside its own boundaries. Thus, fugitive emissions from a new coalmine were Australia's responsibility, and emissions from burning the coal the responsibility of the importing country.

What happens to emissions within Australia would have a larger influence on the global mitigation effort. Emissions within Australia determine whether Australia is contributing its fair share to, or undermining, the global effort within 'concerted unilateral mitigation'. By making the Adani mine emblematic of attitudes to climate change, environmental groups on the one hand and the government on the other succeeded in making the Opposition's position on climate change appear weak or equivocal. This undermined what otherwise would have been an advantage among voters who favoured strong action on climate change. At the same time, the Opposition seeming to equivocate on the Adani mine allowed the government to present itself as the major party that favoured growth in employment and incomes in the coalmining regions, and in the state in which coalmining was proportionately most important.

Several national polls around the time of the election revealed strong and growing support for action on climate change. The Lowy Institute Poll is particularly interesting, as it has asked the same question each year since 2006, allowing the tracking of opinion over time. Respondents are asked to say which of three statements corresponds most closely to their own views:

1. Global warming is a serious and pressing problem. We should be taking steps now even if this involves significant costs.

2. The problem of global warming should be addressed, but its effects will be gradual, so we can deal with the problem gradually by taking steps that are low cost.

3. Until we are sure that warming is really a problem, we should not take any steps that would have economic costs.

I was a director on the board of the Lowy Institute in the early years of the poll and was never comfortable with the questions. I myself would have answered 'yes' to each. I suppose the 'yes' would have been stronger for the second than the third, and strongest for the first. Nevertheless, the questions pick up differences in attitudes to the issue. Changes over time are instructive. In 2006, a percentage in the high 60s favoured the strongest response, compared with only 24 per cent favouring the second. The proportion favouring the first fell steadily, and then more rapidly as Tony Abbott made opposition to action on climate change a feature of his assault on the Malcolm Turnbull Liberal Party leadership, and then on the Labor government. The proportion favouring the first response fell to a low 37 per cent in 2012. It was then well below the proportion favouring the second answer (46 per cent). Since 2012, the proportion favouring the first has increased steadily, to 61 per cent in 2019. The proportion favouring the second has fallen steadily, to 29 per cent in 2019. The third set of respondents include those who think that atmospheric physics is bunkum, plus some who are simply cautious about how much we should pay for mitigation. They numbered 8 per cent in 2006 and swelled to 19 per cent as the leader of the Opposition reached full voice in 2011. They then shrunk steadily to 10 per cent in 2017 and have stayed there since that time. As medicine Nobel laureate Peter Doherty pointed out in his University of Melbourne Festival of Ideas public lecture in 2009, that is a base level of denial that is rarely absent in any area of science with implications for human behaviour.[1]

The Lowy Institute Poll in 2019 revealed a marked elevation in Australian perception of climate change as a threat to national security. For the first time, climate change was first in the list of threats to

Australia's national interests. It ranked ahead of all others, including cyber-attack, international terrorism and North Korea's nuclear weapons program. Climate change was well ahead of China establishing a military base in a South Pacific country.

It would therefore be a brave government that presumed it was immune from electoral reaction if it persisted in a weak response to climate change.

On the long-term changes in opinion tracked by the Lowy Institute Poll, the respondents in the second category are distinguished mainly by concern for the costs of reduced emissions. Awareness is growing that costs in Australia are low and may soon turn into benefits.

Chapter 2 showed that new knowledge in atmospheric physics has reduced uncertainty without greatly changing average expectations. The world made a late and slow start on mitigation, so we lost a few years. Today we have a few less years to get to zero net emissions.

The physical effects of warming are continuing on established unhappy trajectories (average temperatures in Australia; streamflow into Perth dams; inflows into the Murray–Darling system) or tracking above the worst end of the range of anticipated possibilities (average sea-level rise).

Increased scientific knowledge has raised expectations of sea-level rise if mitigation is weak. Increased scientific knowledge of feedback effects suggest that gas concentrations that were thought likely to generate 2°C of warming may bring 2.25°C. Meeting emissions reductions as proposed in recent IPCC reports for 1.5°C is likely to leave us with a 1.75°C rise. That may still underestimate ice–albedo and climate-cycle feedbacks. The scientific work continues.

The balance of international opinion on prudent limits to temperature increases has shifted from 2°C to 1.5°C. There is increased awareness of the insurance value of avoiding the worst outcomes – what I called Type 3 benefits of climate-change mitigation. There is clearer thinking about the ethics of climate change, led by Pope Francis in *Laudato si'* and by thinkers and moral leaders in many other ethical and religious traditions: there is increased awareness of the value of conserving our common planetary home and its biological heritage, and of climate change as a social justice issue.

There have been three other big developments affecting decisions on reducing emissions. One is a huge fall in the cost of equipment for many low-emissions technologies. The second is increased international awareness of the opportunity to remove carbon dioxide in the atmosphere by managing our land in different ways. The third is a decisive fall in the discount rate applied to investment. The latter is especially important because the zero-emissions technologies in energy, transport, industry and land involve mainly capital costs, with low costs of continuing operations. The low costs of capital disproportionately lower the cost of renewable energy and low-emissions industrial and transport technologies in competition with traditional fossil-energy-based technologies.

The economic conditions that gave us the lower discount rate do more than reduce the cost of investing in the zero-emissions economy. They reduce the force of any argument that a loss of material consumption in future is intrinsically less valuable than a loss today. Stagnation in ordinary Australian incomes since the 2011 Review has weakened the case for attributing lower value to a unit of consumption by our grandchildren than by ourselves on the grounds that Australians will be richer in the future.

In the symbolic presentation of Chapter 3, the body of the fish has shrunk and the tail expanded.

There was a major change in the international cooperation regime on climate change after 2008. As it turned out, this improved the prospects for ambitious global mitigation. In Paris, the UNFCCC agreed to hold warming as far as possible below 2°C from pre-industrial levels, and as close as possible to 1.5°C. This is the overriding Paris objective, towards which the national statements of more immediate targets are meant to lead. Each country defines its own immediate national objectives. Unlike earlier attempts to build a 'top-down' binding agreement, Paris was consistent with US political and constitutional realities, China's response to those realities, and with the extension of the international mitigation effort to the major developing countries.

The Paris Agreement is challenged by President Trump's declared intention to withdraw. US withdrawal would be a serious setback. If it were to happen, correction is more likely if the rest of the world continues within the Paris framework. The weight of domestic opinion in favour of effective action is as strong in the United States as in other countries, so there would be continuing political pressure for re-engagement.

So far, the negative example of President Trump has not been directly influential outside the United States – although this would change in a long Trump presidency. More importantly so far, the US–China trade war has reduced economic growth in China below post-2012 expectations. One response has been to start up the engines of the old, fossil-energy economy to offset the downturn. This is a setback for global mitigation.

The strong community interest in combating climate change is universal and growing across the world. Increasing knowledge of the science, the impacts and the ethics are all strengthening popular commitment to action. This is important in authoritarian as well as democratic political systems. There is greater awareness now that many countries are taking strong action, and so less anxiety in each country that others will free-ride on their own contributions.

There is increasing knowledge of a range of co-benefits that accompany action to reduce greenhouse-gas emissions. Lower use of fossil energy reduces air pollution, with measurable effects on health and longevity. This has been especially important in China and increasingly so in the even more heavily polluted cities of India. In countries dependent on imports of fossil energy, again most importantly China, domestic supply of renewable energy has reduced import prices for coal, oil and natural gas and reduced vulnerability to disruption of imports. For countries that are potential suppliers of the capital goods of the low-carbon world economy – including Germany, China, Japan, Korea and the United States – there is an advantage to moving early. For Australia, there is the potential to be the world's main exporter of energy-intensive or carbon-intensive manufactures in the zero-carbon world. Co-benefits of capturing more carbon in Australian soils and landscapes include savings on chemical fertilisers, greater tolerance of lower rainfall and, on average over time, higher productivity.

Holding temperature increases to 1.5°C needs zero net emissions in the world within three decades. We have started slowly. Continue slowly for long, and the world will have to reach zero emissions much earlier than mid-century.

Both the UNFCCC and commonly accepted ethical principles tell us that developed countries should move to zero net emissions earlier than developing ones. Net zero emissions in Australia before mid-century is implausible if we have only reduced emissions by 26 per cent from 2005 levels by 2030. It is unrealistic to expect the government to revise the target, endorsed by electoral victory, in the current parliamentary term. It may not be likely that the government will revise the target soon after that – especially with the current emissions trajectory being well above what is necessary to meet the formal target. However, Australia could do much better than the announced target if we put in place a few policies to encourage emission reductions in some areas where they can be achieved at low or negative cost. None of these policies would breach explicit government election commitments. Demonstration that much larger reductions are possible alongside strong expansion of economic activity will create an environment in which the acceleration of progress is not only feasible, but electorally attractive. Properly explained, the attraction will be strongest in rural and provincial areas and in Queensland north of Brisbane, where the Coalition won its House of Representatives majority in May 2019.

A new framework

How do we get started towards zero net emissions?

At times when national governments have not acted on climate change, state governments have had major impacts in the three developed countries with the highest per capita emissions – Australia, the United States and Canada.

Other countries' actions contribute to falling costs of reducing emissions in Australia. Cheaper solar panels, wind turbines, batteries and electric cars, driven by other countries' climate and energy policies, are leading to reduced emissions here.

Actions by individuals are part of the story of the reduced Australian use of electricity over the past decade. It is part of the story of many Australians' early adoption of solar PV and now of batteries and electric vehicles.

So there will be some progress in reducing emissions independently of national policy. But coherent national policy is necessary to achieve strong climate objectives.

A decade ago, I saw economic and climate change goals as needing to be closely aligned. My recommendations had as hard an eye on preserving the strengths of the Australian economy as on certainty in achieving climate change objectives. As I said in the concluding pages of the 2011 Review:

[R]eductions of emissions under the market-based scheme proposed in this book ... will come from everywhere. Consumers will use less energy and other goods and services that embody high levels of emissions. Natural gas exporters will try harder to find opportunities for sequestration of fugitive emissions and the wastes from liquefaction. Landowners will think hard about the parts of their properties that would have more value as carbon sinks than carrying sheep. Lots of people with clever ideas that reduce emissions will find equity investors and lenders more interested than they were before … Once we put the carbon pricing incentives in place, millions of Australians will set to work finding cheaper ways of meeting their requirements and servicing markets. We don't know in advance what the successful ideas will be, but I'm pretty sure that there will be extraordinary developments in technology … That is the genius of the market economy … And that is why reliance on regulatory approaches and direct action for reducing carbon emissions is likely to be immensely more expensive than a market approach … The really big cost [of the regulatory approach] would be the entrenchment of the old political culture that has again asserted itself after the [late] 20th century period of reform.[2]

Comprehensive carbon pricing is the centrepiece of any environmentally and economically efficient program to reduce emissions. The emissions trading scheme operating in Australia from mid-2012 to mid-2014 was an exemplar of that. Among much else, it would have allowed unlimited entry into the European Union's high-priced emissions trading scheme for Australian exports of energy-intensive metals. It would have brought Australia, with its strong interest in legitimate land-based credits, and Europe, with its history of doubting their legitimacy, into a dialogue on how to make carbon farming work in an environmentally strong trading scheme. But the politics of the past decade have for the time being poisoned the well for carbon pricing. So how do we get to net zero emissions before 2050 without drinking from this poisoned well?

Carbon pricing was not the whole story of incentives to reduce emissions when it was operating. Other elements of the Clean Energy Future, preserved by the Senate in 2014, play important roles. One is fiscal support for innovation. The Australian Renewable Energy Agency (ARENA) has performed that role for electricity supply, but not generally for innovation in the new zero-emissions industrial and land-management technologies. ARENA has recently given priority to use of renewable energy in minerals processing. Public funding for research and development, with a strong focus on low-cost carbon measurement technologies, needs to be made available to sequestration of carbon in soils, pastures, woodlands, forests and plantations.

Uncertainty in policy can raise the supply price of investment, and so slow and raise the cost of transition. The Clean Energy Finance Corporation (CEFC) has played an important role in correcting that market failure for renewable energy, especially for debt. Its role could be extended to industrial applications of clean energy.

The Climate Change Authority (CCA) for a while performed the role for which it was designed: to advise the parliament authoritatively and independently of partisan politics on the emissions-reduction targets that would meet national and global climate goals. Its strength could be restored.

The Clean Energy Regulator continued to manage offsets from the Carbon Farming Initiative after the abolition of carbon pricing, with funding under the Emissions Reductions Fund (ERF). The ERF contains a system of baselines, with the large emitters of greenhouse gases purchasing and surrendering ACCUs to balance any emissions above the baselines. The baselines are undemanding, and not designed to push down emissions. New producers – for example, a new coalmine or a new LNG processing facility – are given baselines based on established industry experience, Nevertheless, in 2016–17, this led to the purchase of about 450,000 ACCUs, and in the following year around half of that amount. Some large emitters have postponed liability by applying for multi-year averaging of their performance against baselines and are likely to be in the market for ACCUs at the end of the averaging period.

While the constraints are weak, there is an established administrative structure that could be brought into play to meet more demanding emissions constraints. For example, the sources of fugitive emissions covered by the ERF safeguards mechanism could have the baseline phasing down to zero over the period to 2030. The Clean Energy Regulator can also certify legitimate carbon sequestration as ACCUs for sale into the voluntary market, or into an international emissions trading system in which Australia participates.

The machinery for administering carbon pricing through the Clean Energy Regulator was never dismantled and could be brought back into operation much more quickly than it was installed at the beginning. So a government committed to reducing emissions consistently with the Paris Agreement would not be starting from scratch.

While there would be undoubted benefits in stable, comprehensive carbon pricing, I would not support going back to such a system without unequivocal bipartisan support in the Australian parliament. There are two reasons. One is that success with carbon pricing depends on confidence in policy durability. With uncertainty about the longevity and stability of the carbon price, the economy would pay a cost without receiving a corresponding benefit through reduction of emissions. Neither a grant nor a guarantee of a revenue stream has that shortcoming. Better to use other instruments until we are confident that the poison in the carbon pricing well has diluted to non-lethal concentrations.

But we should remember that large economic and environmental gains will accompany a return to comprehensive carbon pricing at some time in the possibly distant future.

In the meantime, we can make a lot of ground with measures of other kinds – assisted by reductions in the cost of low-emissions technologies. I suggest the following guide to new incentives:

1. Use established institutions with transparent processes as much as possible. Extend the roles:
    › of ARENA to support innovation across the range of low-emissions technologies;
    › of the CEFC to cover extended versions of the ACCC's Recommendation 4;
    › of the Clean Energy Regulator to administer an expanded offsets system;
    › of a restored CCA for independent advice on emissions-reduction targets;
    › of the Productivity Commission to review safeguard arrangements and, if carbon pricing is reintroduced in future, exemptions and special treatment of trade-exposed industries.

2. Where new incentives are introduced, make them of general application without discretion or negotiation in their application. Confine any consultation with business to the principles underlying policy and keep it away from applications affecting individual enterprises.

I would suggest a limited initial expansion of incentives to reduce emissions:

3. For electricity,
    › application of the ACCC's Recommendation 4, available to all companies meeting clearly defined criteria;
    › rewarding of private investors in unregulated transmission for the value of services provided to the regulated system;

> maintenance of system security under rules developed by AEMO over the past two years;

> underwriting reliability with increasing use of intermittent renewable energy through the Snowy National Energy Guarantee proposed in Chapter 4.

4. For industry,

> increased funding for ARENA and the CEFC to expand coverage from renewable energy to low-emissions technologies more generally, including the promotion of the hydrogen economy and the biomass-based chemical industry.

5. For transport,

> development of principles for Commonwealth, state and local government participation in provision of electric-vehicle charging systems, followed by substantial public fiscal support for infrastructure for charging electric vehicles.

> development of power pricing rules that ensure that electric vehicles reduce rather than increase power distribution costs.

6. For fugitive emissions,

> require the purchase of offsets (ACCUs) for any emissions that exceed a baseline that falls to zero through the 2020s. If this is not taken up by the Commonwealth, it could be implemented by one, some or all state governments under environmental and mineral leasing powers.

7. For the land sector,

> substantial increase in research funding to define potential and mechanisms for increasing sequestration of carbon in the Australian landscape;

> commitment of substantial funding for research, development and commercialisation of low-cost and reliable measurement of carbon sequestration in soils, pastures, woodlands, forests and plantations;

> development of the ERF into a general rather than auction-based system for rewarding genuine carbon sequestration in agriculture and land-use change (including geological BECCS); a minimum objective to secure a market for all valid offsets generated within the land sector;

> facilitation of sale of offsets for exceedance of safeguard baselines for fugitive and other emissions and into the voluntary market;

> initial ERF funding from making the Commonwealth Government's commitment to $2 billion for the Climate Solutions Fund available for immediate use until exhaustion;

> working with the policy leaders of the European and other carbon markets to agree conditions and rules under which legitimate ACCUs could be sold into the markets, whether or not Australia itself has an emissions trading scheme;

> enforcement of intelligently framed controls on landclearing.

We don't have to get to our policy endpoints in one step.

These suggestions would put us on a path the extension of which would lead to 50 per cent reduction of Australian emissions (on 2005 levels) by 2030. That would be a credible launching position for zero net emissions before 2050.

Electricity emissions would fall below 50 per cent of 2005 levels on the back of the anticipated expansion of renewable energy. Land sector emissions, already negative, would fall further, without crediting reductions used to offset emissions in other sectors or countries or in the voluntary markets. Transport emissions would decelerate and begin a slow descent as the public transport share of Australian urban transport continues to rise and Australia begins to make use of opportunities nurtured in other countries for low-cost electrification of road transport. Agricultural emissions would be substantially below 2005 levels as deterioration of climate continues to constrain cattle and sheep numbers, and with slow take-up of other abatement opportunities. Fugitive emissions would fall to zero as gas and coal suppliers take advantage of relatively low-cost means of reducing them, as they did between 2008 and 2014, and offset remaining emissions. Industrial output would expand rapidly without any increase and probably with some decline in emissions.

Far from raising electricity prices and reducing economic activity, the measures listed here would lead to substantially lower wholesale prices to major industrial users of power. This would then help lower prices for other users.

These early measures would nurture the use of the four great opportunities for Australian industrial leadership in the post-carbon world economy: globally competitive renewable power; use of competitive electricity and hydrogen for local processing of a high proportion of Australian mineral production; an abundance of biomass for the chemical manufacturing industries; and low-cost biological and geological sequestration of carbon wastes.

Australia as the superpower of the post-carbon world economy

In sketching the possibilities for Australia in 2011, I emphasised one condition for future success:

> We can be sure that if we are working within a strong global mitigation scenario, coal soon and natural gas later will cease to be competitive everywhere unless economically efficient sequestration technologies have emerged.
>
> That does not mean we lose our advantage in energy-intensive industries ... our natural advantages in a wide range of

low-emissions energy sources are likely to keep us competitive ...
[but for] Australia to remain an internationally competitive location for energy-intensive industries in a world of strong, global
carbon constraints ... [w]e will need to be seen as one of the countries that is focussing on the industries of the future and not only
on protecting the industries of the past.[3]

I would put the opportunity much more positively today. I discussed in Chapter 4 how secure, reliable power can be made available
in the best Australian locations at globally competitive prices.

Australia is by far the world's largest exporter of mineral ores for
making iron and aluminium. In the zero-carbon future world economy,
we are economically best positioned to turn these minerals into iron
and aluminium metal. We are naturally a competitive supplier at least
to Asian markets of electricity-based industrial goods more generally.

I discussed in Chapter 5 how biomass will be scarce and valuable in a world where emissions from coal, gas and oil have made
them unacceptable sources of raw materials for petrochemicals. Australia's exceptional endowment of forests and woodlands per person
gives it a comparative advantage in biological raw materials for industrial processes. Careful management of our land in general and our
vast semi-arid regions in particular will unlock a rich resource for
biosequestration.

The world will come to value highly opportunities for negative
emissions. We can use the oldest, best-established and lowest-cost technology for negative emissions: photosynthesis, followed by the capture
of carbon in the landscape.

A fourth natural source of Australian industrial strength in the post-
carbon world economy will be geological sequestration. Combined
with biomass-based industrial activites, this will expand opportunities
for negative emissions.

Big questions remain. The answers will appear before us as millions of Australians respond to opportunity, shaped by policy as it evolves. But the broad outlines are clear now.

In a zero-carbon world economy, there would be no economic sense in any aluminium or iron smelting in Japan or Korea, not much in Indonesia, and enough to cover only a modest part of domestic demand in China and India. The European commitment to early achievement of zero net emissions opens a large opportunity there as well. How big in Australia? Converting one-quarter of Australian iron oxide and half of aluminium oxide exports to metal would add more value and jobs than current coal and gas combined. That is before the ammonia, the silicon, the other processed minerals, the chemical manufactures and the carbon farming.

Where? I highlighted in Chapter 5 that there will be special advantages at first in the old transmissions nodes for coal generation: the Upper Spencer Gulf, Collie, the Latrobe Valley, Newcastle, Gladstone. The transmission systems that took coal-based power to the cities and other centres of big demand will find new value bringing in renewable energy. Old industrial towns have legacies of infrastructure and industrial culture that have value for new industrial activity. Northern Tasmania and Portland present variations on the theme. Over time, other regions will assert themselves. Regions without the transmission legacy will need to break free from the regulatory shackles that make old energy expensive.

How soon? Gradually, then suddenly. Some change is ready to happen now. With globally competitive prices for electricity, the economically logical place for new aluminium smelting when the world needs more is adjacent to the alumina production in the southwest of Western Australia. In recent years, high energy prices have pointed to closure of established aluminium smelters in Portland, Newcastle and Gladstone. With globally competitive power, they are candidates for revival and expansion. With globally competitive power, Australia becomes the natural locus for supply of the world's immense increases in demand for pure silicon. The processing of many minerals required in increasing proportions by the post-carbon world economy fits naturally here – lithium, titanium, vanadium, nickel, cobalt, copper. So does energy-intensive processing of carbon fibre and other energy-intensive processes drawn from biomass. Renewable carbon fibre will have a non-renewable Australian base from which to build.

Some new zero-emissions industrial products will require no materials other than globally competitive power. Ammonia, electricity transmitted by high-voltage direct-current cable to and through Indonesia and Singapore to the Asian mainland and exportable hydrogen are among them. Movement will come gradually, initially with public support for innovation. Then suddenly, as business and government leaders come to realise the magnitude of the Australian opportunity, and as humanity enters the last rush to prevent being overwhelmed by the rising costs of climate change. The pace will be governed by progress in decarbonisation globally. That calibration will suit us, as the new strengths in the zero-carbon world economy grow with the retreat of the old.

*

'An old dog for a hard road', I wrote in the letter with my first Review delivered to the prime minister and premiers in September 2008.

Eleven years later and we are still on the road.

I was in Central Queensland in April with two grandchildren, Kai and Harvey, where Clancy of the Overflow went droving, Banjo wrote 'Waltzing Matilda', the Queensland and Northern Territory Aerial Services built and flew the planes, and the shearers gathered beneath the Tree of Knowledge to form the Australian Labor Party.

It overlaps Adani country, where Clive Palmer and Gina Rinehart want to build new coalmines. This is the best solar country in eastern Australia and just about as good as any in the world. Its sun will fuel Gladstone when it supplies Japan and Korea with hydrogen for their cars and power stations, and China with silicon for its computers and solar panels; and much more of northeast Asia's aluminium than now.

The rain has come at last, in a great dump. The sun is glancing from the shin-deep Mitchell grass as far as an old eye can see. The Thomson River is testing its banks at Longreach, swirling like a homing pigeon, orienting itself for the long, slow glide down the Cooper to Lake Eyre.

I will leave the last word to 'Clancy of the Overflow':

Yes, we saw the vision splendid,
of the sunlit plains extended,
and at night the wondrous glory
of the everlasting stars.

ACKNOWLEDGEMENTS

This book is based on a series of six public lectures held at the University of Melbourne in April and May 2019: 'The Climate and Energy Transition in Australia'. The lectures were hosted by the Faculty of Business & Economics (my home base within the university, where I am a Professorial Research Fellow in Economics), The Melbourne Energy Institute (where I am a Distinguished Fellow and a member of the Advisory Board), The Australian German Energy Transition Hub (where I chair the International Advisory Board) and the Melbourne Sustainable Society Institute. Video recordings of the lectures are accessible on the websites of these institutions. I am grateful to Paul Kofman, Michael Brear, Malte Meinshausen, Rebecca Burdon and Brendan Gleeson for hosting the lecture series, and for hosting my continuing education on the subject matter of the book.

My long-time friend and colleague Max Corden insisted that I write this book, and present more simply the things that I said in the Garnaut Climate Change Review 2008 and Australia in the Global Response to Climate Change 2011.

On the atmospheric physics and other science, I am especially grateful to Professors Will Steffan, David Karoly, Mike Sandiford and Malte Meinshausen. Zebedee Nicholls went beyond my general education and prepared charts for the lectures.

On land use sequestration, Kate Dooley brought me up to date with the literature and provided rich insights into recent developments in thought and action.

On energy markets including preparation of visual material, I am deeply indebted to Dylan McConnell. Dylan keeps us all up to date and on our toes.

Michael Lord's excellent work for Beyond Zero Emissions on Australia's industrial options has complemented my own through this year and is having an important impact on the Australian discussion. This follows the earlier work of Gerard Drew for Beyond Zero Emissions on Australia as an energy superpower. Leslie Martin's work on how consumer electricity markets work enriched the discussion in the electricity lecture.

This book has grown from my two climate change Reviews. I was fortunate in the excellent teams who worked with me. The first was led by Ron Ben David and the second by Steven Kennedy, both of whom contributed analytic insights that did much to shape the reports and their recommendations. That the basic economic framework has stood the test of time I owe much to my old teams and especially Stephen Howes and Frank Jotzo. More generally, there were superb teams of clever and well informed mostly young professionals from many fields, heavily committed to our presumptuous task. Most of us keep in touch as we maintain interests in old big things. I made full acknowledgements in the books from each Review and draw attention again to those earlier expressions of thanks.

I have learned a great deal with many other people as we explored together over many years the issues raised in this book; none more than Reuben Finighan on the appendix to Chapter 3 and Robin Batterham and Alan Finkel on the contents of Chapters 4, 5 and 7.

I have drawn inspiration and learned a great deal over these past eleven years from many people outside the big Australian cities.

I mention in particular the leaders of thought and action on climate change and the related economic opportunity in the Upper Spencer Gulf and Eyre Peninsula of South Australia (where I have visited more than a dozen times to talk about climate change and the energy opportunity), the Latrobe Valley, Ballarat, Shepparton, Swan Hill, Mildura, Geelong, Albury, St Clements Galong, Barcaldine, Longreach, Gladstone, Winton, Hughenden, Collie, the Western Australian wheatbelt and east Arnhem Land.

In putting together the lectures and the book, Susannah Powell has been a consistent contributor and supporter. Thanks also for excellent work on the book to Loulou Gebbie, Kirstie Innes-Will, Julia Carlomagno, Dion Kagan and for the strategic eye and pen of Chris Feik.

Ross Garnaut
University of Melbourne
1 October 2019

NOTES AND SOURCES

Chapter 1

1   W.J. Young et al., 'Science Review of the Estimation of an Environmentally Sustainable Level of Take for the Murray–Darling Basin', Final report to the Murray–Darling Basin Authority, CSIRO, November 2011, p. 9.

2   Murray–Darling Basin Authority, 'Climate Change and the Murray–Darling Basin Plan', MDBA discussion paper, February 2019, p. 14.

3   Here, recommendation four of the Australian Competition and Consumer Commission's 2018 report on the electricity sector would provide the necessary support. See Australian Competition and Consumer Commission (ACCC), 'Restoring Electricity Affordability and Australia's Competitive Advantage', Retail Electricity Pricing Inquiry final report, Australian Competition and Consumer Commission, June 2018, Chapter 2.

Chapter 2

1   Lindsay Lenhe at Mildura assisted me with putting together historical data from MDBA sources.

2   Pope Francis (Jorge Mario Bergoglio), *Encyclical Letter Laudato si' of the Holy Father Francis: On Care for Our Common Home*, 24 May 2015, p. 21.

3   Genesis 1:28.

4   Pope Francis, *Laudato si'*, pp. 48–9.

5   Ross Garnaut, *Dog Days: Australia After the Boom*, Black Inc., Melbourne, 2013, p. 225.

6   Ross Garnaut, *Dog Days: Australia After the Boom*, Black Inc., Melbourne, 2013, p. 211.

Chapter 3

1   William R. Cline, *The Economics of Global Warming*, Columbia University Press, New York, 1992; William Nordhaus, *Managing the Global Commons:*

The Economics of Climate Change, MIT Press, Cambridge, 1994; William Nordhaus, *A Question of Balance: Weighing the Options on Global Warming Policies*, Yale University Press, New Haven, 2008; Nicolas Stern, *The Economics of Climate Change: The Stern Review*, MIT Press, Cambridge, 2007.

Chapter 4

1   These upward pressures on prices and proposals for correcting them were the catalyst of a review by the Australian Consumer and Competitor Commission in 2018.

2   Ross Garnaut, *The Garnaut Review 2011: Australia in the Global Response to Climate Change*, Commonwealth of Australia, Canberra, 2011 and Cambridge University Press, Port Melbourne, 2011, p. 154.

3   ACCC, 'Restoring Electricity Affordability and Australia's Competitive Advantage', p. 88.

4   Ibid., p. 17.

5   See 'ANU finds 22,000 potential pumped hydro sites in Australia', Energy Change Institute, Australian National University, Canberra, 2017; Andrew Blakers et al., 'Global Pumped Hydro Atlas', ANU College of Engineering and Computer Science, Australian National University, Canberra, 2017.

6   Alan Finkel, 'Independent Review into the Future Security of the National Electricity Market – Blueprint for the Future', Commonwealth of Australia, Canberra, 2017.

Chapter 5

1   Energy Transition Commission, *Mission Impossible, Reaching Net Zero Carbon Emissions from Harder to Abate Sectors by Mid-Century*, London, 2019.

Chapter 6

1   Marcus Brazil, 'Australia's Electricity Grid Can Easily Support Electric Cars – If We Get Smart', *The Conversation*, 12 April 2019.

2   David P. Byrne, Andrea La Nauze and Leslie A. Martin, 'Tell Me Something I Don't Already Know: Informedness and the Impact of Information Programmes', *The Review of Economics and Statistics*, vol. 100, no. 3, July 2018, pp. 510–27.

Chapter 7

1   Michael Battaglia, 'Greenhouse Gas Mitigation: Sources and Sinks in Agriculture and Forestry' in Helen Cleugh et al. (eds), *Climate Change: Science and Solutions for Australia*, CSIRO Publishing, Collingwood, 2011, p. 97.

2   Bronson W. Griscom et al., 'Natural Climate Solutions', Proceedings of the National Academy of Sciences of the United States of America, vol. 114, no. 44, 2017, www.pnas.org/content/114/44/11645

3   Kate Dooley and Sivan Kartha, 'Land-based Negative Emissions: Risks for Climate Mitigation and Impacts on Sustainable Development', *International Environmental Agreements: Politics, Law and Economics*, vol. 18, no. 1, pp. 79–98.

4   Malte Meinshausen and Kate Dooley have shown that we can get most of the way to the Paris goals through renewables-based electrification and the land use transformation. See their chapter 'Mitigation Scenarios for Non-energy GHG' in Sven Teske (ed.), *Achieving the Paris Climate Agreement Goals: Global and Regional 100% Renewable Energy Scenarios with Non-energy GHG Pathways for +1.5°C and +2°C*, Springer Publishing, New York, 2019.

5   Colin Clark, *The Conditions of Economic Progress* (3rd edition), Macmillan, London, 1957.

6   Ibid., pp. 488–9.

7   The Lancet Commissions, 'Food in the Anthropocene: The EAT-Lancet Commission on Food, Planet, Health', *The Lancet*, vol. 393, no. 10170, February 2019, pp. 447–92.

8   Intergovernmental Science-Policy Platform on Biodiversity and Ecosystem Sciences, *Global Assessment Report 2019*, IPBES, 2019.

9   Paula J. Peeters and Don W. Butler, 'Brigalow – Regrowth Benefits Management Guideline', Department of Science, Information Technology, Innovation and the Arts, Brisbane, 2014.

10 Cited in ibid.

11 There is now a considerable literature on soil carbon. The rough arithmetic says that 2.4 trillion tonnes of terrestrial carbon in the top 2m of the planet's soils hold four times as much carbon as all the world's plants. In the atmosphere, there is now approaching a trillion tonnes of carbon

embedded in carbon dioxide. An article in *The Conversation* provides an introduction: Budiman Minasny, Alex McBratney, Brendan Malone, Uta Stockman, 'Eyes Down: how setting our sights on soil could help save the climate', *The Conversation*, 1 December 2015. See also W.C. Clark, *Carbon Dioxide Review*, Oxford University Press, New York, 1982; N.H. Batjes, 'Total carbon and nitrogen in the soils of the world', *European Journal of Soil Science*, vol. 47, June 1996, pp. 151–163.

12 Zhongkui Luo, E. Wang, O. Sun, 'Soil Carbon Change and Its Responses to Agricultural Practices in Australian Agro-ecosystems: A Review and Synthesis', *Geoderma* 155, 2010, pp. 211–23; Rachel Meyer, B.R. Cullen, R.J. Eckard, 'Modelling the Influence of Soil Carbon on Net Greenhouse Gas Emissions from Grazed Pastures', *Animal Production Science* 56, p. 585–93; P. Ciais, C. Sabine, G. Bala, L. Bopp, V. Brovkin, J. Canadell, A. Chhabra, R. DeFries, J. Galloway, M. Heimann, C. Jones, C. Le Quéré, R.B. Myneni, S. Piao and P. Thornton, 'Carbon and Other Biogeochemical Cycles', in *Climate Change 2013: The Physical Science Basis. Contribution of Working Group I to the Fifth Assessment Report of the Intergovernmental Panel on Climate Change*, Cambridge University Press, Cambridge, United Kingdom and New York; Shu Kee Lam, D. Chen, A.R. Mosier, R. Roush, 'The Potential for Carbon Sequestration in Australian Agricultural Soils Is Technically and Economically Limited', *Scientific Reports*, vol. 3, no. 2197, 2013; Pete Smith, et al., 'Biophysical and Economic Limits to Negative CO2 Emissions, *Nature Climate Change* 6, 2016, pp. 42–50; Emanuele Lugato, A. Leip, A. Jones, 'Mitigation Potential of Soil Carbon Management Overestimated by Neglecting N2O Emissions', *Nature Climate Change*, vol. 8, no. 3, pp. 219–23; Budiman Minasny et al., 'Soil Carbon 4 Per Mille', *Geoderma* 292, 2017, pp. 59–86.

13 Lam et al., 2013; Lugato et al. 2018; Robert E. White, et al., 'A Critique of the Paper "Soil Carbon 4 Per Mille" by Minasny et al. (2017)', *Geoderma* 309, pp. 115–17.

14 Charles Massy, *Call of the Reed Warbler: A New Agriculture – A New Earth*, University of Queensland Press, St Lucia, 2017.

15 Pablo S. Alvarez-Hess, S.M. Little, P.J. Moate, J.L. Jacobs, K.A. Beauchemin, R.J. Eckard, 'A Partial Life Cycle Assessment of the Greenhouse Gas Mitigation Potential of Feeding 3-Nitrooxypropanol and Nitrate to Cattle', *Agricultural Systems*, vol. 169, 2019, pp. 14–23.

16 Shyama Ratnasari and Jayatilleke Bandara, 'Changing Patterns of Meat Consumption and Greenhouse Gas Emissions in Australia: Will Kangaroo Meat Make a Difference?' *PLOS ONE*, vol. 12, no. 2, 2017.

17   Damien R. Farine, et al., 'An Assessment of Biomass for Bioelectricity and Biofuel, and for Greenhouse Gas Emission Reduction in Australia', *Global Change Biology Bioenergy*, vol. 4, no. 2, 2012, pp. 148–75.

18   United Nations Framework Convention on Climate Change (UNFCCC), 'Paris Agreement', 2015, Article 4.1.

19   Kate Dooley and Aarti Gupta, 'Governing by Expertise: The Contested Politics of (Accounting For) Land-based Mitigation in a New Climate Agreement', *International Environmental Agreements: Politics, Law and Economics*, vol. 17, no. 4, August 2017, pp. 483–500.

20   UNFCCC, 'Paris Agreement', 2015.

21   Ibid.

22   House of Representatives, 'Carbon Farming Initiative Amendment Bill 2014', Parliament of Australia, Canberra, 2014.

Chapter 8

1   See also Peter Doherty, *The Knowledge Wars*, Melbourne University Press, Melbourne, 2015.

2   Garnaut, 2011, pp. 173–4.

3   Ibid., p. 172.

INDEX

Ross Garnaut is Professorial Research Fellow in Economics at the University of Melbourne. In 2008, he produced the Garnaut Climate Change Review for the Australian government. He is the author of many books, including the bestselling *Dog Days* (2013).

Lightning Source UK Ltd.
Milton Keynes UK
UKHW041243220321
380778UK00006B/1424